GENERATION HOPE

Published in the UK by Scholastic Children's Books, 2020
Euston House, 24 Eversholt Street, London, NW1 1DB
A division of Scholastic Limited

London – New York – Toronto – Sydney – Auckland
Mexico City – New Delhi – Hong Kong

ISBN 978 14071 9993 1

A CIP catalogue record for this book is available from the British Library.

Printed in Italy by L.E.G.O S.p.A
Papers used by Scholastic Children's Books are made from wood grown in sustainable forests.

1 3 5 7 9 10 8 6 4 2

www.scholastic.co.uk

GENERATION HOPE

Kimberlie Hamilton

Character illustrations
by Risa Rodil

SCHOLASTIC

CONTENTS

GENERATION HOPE

ACTIVIST KIDS MAKING A DIFFERENCE

WANT TO MAKE A DIFFERENCE?

People say that young people are the future, but kids these days aren't waiting to grow up to take action. They are raising awareness, fighting back and inspiring positive change. They are gearing up like never before. **They've had enough!**

This book features young activists around the world – all under twenty years old – who are working hard to fix the planet and help the people who live on it. Some do big things, others do smaller things, but every one of these rebels with a cause is a force for good. Added bonus: dozens and dozens of ideas for what you can do to make the world a better place too!

Ready to get started? Just find one cause that ignites your fire and go from there. No matter the obstacles, no matter your age, **YOU** have the potential to do something amazing.

THE FUTURE IS NOW!

IMAGINE IF... EVERY KID WHO READS THIS BOOK TAKES ACTION?

JUST THINK WHAT YOUR GENERATION COULD ACCOMPLISH!

CLIMATE CHANGE-MAKERS

Spoiler alert: climate change is real. When people talk about climate change, they are talking about what happens when the earth gets hotter. And it's heating up faster than ever before.

All this heat is messing with the earth's natural balance. It's causing more weird weather, like hurricanes, wildfires, droughts and rainstorms. It warms the oceans and melts the ice caps at both poles. No matter where we live, climate change affects us.

Climate change is complicated, but the end result is simple. There isn't much time left to put on the brakes and stop destroying the planet. Yikes!

That's the bad news. **The good news is that activists are joining together and demanding action.**

ACTIVISTS LIKE SWEDISH TEENAGER GRETA THUNBERG.

When Greta first heard about climate change, she took it hard. She felt angry, depressed and frustrated. Why wasn't anyone else freaking out about it? She had to do something. So, she skipped school one Friday and sat outside the Swedish parliament with a sign that said 'School Strike for Climate'.

This solo act of rebellion launched an international youth movement, **#FridaysForFuture.** Greta went from schoolgirl to eco-hero in less than a year.

Greta now speaks at big conferences and tells world leaders what a terrible job they're doing. She's published a book and met VIPs like the Pope and Barack Obama. She's been on the covers of *Time* and *Vogue*. Oh, and one more thing … she's the youngest person **EVER** to be nominated for the Nobel Peace Prize.

In a word, Greta is **HUGE**. Although some people still think climate change is a joke, she doesn't let that stop her. She's on a mission to save the planet.

Is that your cause too? If so, keep reading!

The Greta Generation

In March 2019, more than 1.5 million students took to the streets worldwide to strike for the climate. Six months later, four million people took part in a global strike. It was one of the biggest protests in history!

"TO DO YOUR BEST IS NO LONGER GOOD ENOUGH. WE MUST ALL DO THE SEEMINGLY IMPOSSIBLE."

GRETA THUNBERG

KIDS TAKING ACTION

ZEL WHITING

Born: 2006 **From:** Australia

Known for: organizing student strikes

"I DON'T BLAME ALL ADULTS. THE TRUTH HAS BEEN KEPT FROM THEM. BUT WE NOW KNOW WHAT'S HAPPENING."

The summer of 2018 was one of the hottest on record. Temperatures 'Down Under' reached 49.5°C along the coast. For Zel, that was a real wake-up call about global warming. So was the fact that half of the Great Barrier Reef is dead.

Zel felt he had no choice but to get involved. He now helps organize student strikes and protests against things like fossil fuel projects. Zel hopes his country will use 100 per cent renewable energy by 2030. That's a huge challenge in Australia. The prime minister doesn't support the strikes, but Zel has no intention of stopping. After all, his future – and the future of the planet – is at stake.

JAMIE MARGOLIN

Born: 2001 **From:** USA

Known for: founding Zero Hour

"WE WON'T STOP FIGHTING UNTIL WE GET THE CHANGE WE NEED AND DESERVE."

When wildfires in Canada left Seattle blanketed in smog, experts blamed climate change. Jamie had a sore throat and headaches for weeks. She finally realized that elected leaders weren't going to do anything. It would be up to her generation to take action.

Jamie started her own environmental group, Zero Hour, and along with twelve other kids sued the state of Washington. They said politicians' lack of action was putting kids' health and happiness at risk. The lawsuit was dismissed but Zero Hour now has twenty-five local branches planning youth climate marches all over the world!

TIMOCI NAULUSALA

Born: 2005 **From:** Fiji

Known for: speaking out at a UN conference

"CLIMATE CHANGE IS HERE TO STAY UNLESS YOU DO SOMETHING ABOUT IT."

Timoci was devastated when his village was wiped out by a cyclone. His home, his school, his source of food, water, money … everything was gone. He wrote an essay about it and was invited to speak at a United Nations (UN) conference. The audience included hundreds of VIPs, including twenty-five world leaders. No pressure!

Dressed in a bright tropical shirt, Timoci stood before the crowd and began to speak. He was a small boy with a big smile and an even bigger message. He said that speeches wouldn't fix climate change. The blaming-and-waiting game was over and leaders needed to do more – a LOT more. As the crowd applauded, he shook hands with some of the world's most powerful people!

ANUNA DE WEVER

Born: 2001 **From:** Belgium

Known for: organizing the first climate strike in Belgium

"WE ARE THE LAST GENERATION TO ACTUALLY BE ABLE TO SAVE THIS WORLD."

When Anuna found out that leaders weren't doing anything to stop global warming, she was so angry that she helped to organize Belgium's first climate strike. Three thousand people showed up. The next week there were 15,000 and then 35,000. The strikes grew and became part of a mass global movement.

Since then, Anuna has been interviewed on the news, given a TED Talk and co-authored a book. She's also met with politicians, hoping to wake them up about climate change. In her words, they all act like children, hoping that the crisis will somehow go away.

IT'S NOT T
SAVE TH

Global warming is largely caused by human activities, such as farming and chopping down forests and burning fossil fuels such as coal, oil and gas. All these things allow harmful amounts of carbon dioxide (CO_2) to build up. This 'greenhouse gas' acts like a blanket wrapped around the earth.

Sounds pretty bad, huh? Well, it IS pretty bad. But there's hope. Experts believe we can still stop the terrible damage being done to the planet. There are ways to cut our emissions in half over the next ten years, like using more wind and solar power and not cutting down trees.

Flygskam is a Swedish word that means 'flying shame'. One plane ride can cancel out twenty years of recycling! That's why more and more people are choosing to travel by train.

"WE DESERVE A SAFE FUTURE.
AND WE DEMAND A SAFE FUTURE.
IS THAT REALLY TOO MUCH TO ASK?"

GRETA THUNBERG

O LATE TO PLANET!

WE MUST ALL MAKE CHANGES TO HALT CLIMATE CHANGE. HERE ARE 20 THINGS TO START DOING TODAY:

1. Unplug electronics when not in use.
2. Turn off lights whenever you're not using them.
3. Take short showers instead of baths.
4. Collect rainwater to water your garden and houseplants.
5. Plant a tree to help absorb CO_2.
6. Wrap gifts in newspaper or re-use wrapping paper.
7. Carry a reusable water bottle wherever you go.
8. Give away unwanted items instead of tossing out.
9. Wear a jumper instead of cranking up the heating – and open a window instead of using air con or electric fans.
10. Avoid anything made with palm oil, as producing it destroys animal habitats!
11. Eat more vegetables and plant-based sources of protein.
12. Eat foods that are organic, in-season and locally grown.
13. Avoid using anything with chemicals that can harm the environment.
14. Sign up for community clean-up projects.
15. Walk or cycle as much as possible.
16. Use public transport instead of taking the car.
17. Steer clear of plastic such as bags, straws, packaging and utensils.
18. Ask your parents to buy sustainably sourced fish.
19. Get creative with leftovers instead of throwing them away.
20. Create a compost heap in your garden.

ANIMAL ADVOCATES

There are some people out there who think animals are property that don't need to be treated with kindness or respect. Wrong! **Being mean to animals is NEVER okay** and activist kids are doing something about that.

DO YOU BELIEVE THAT ANIMALS . . .

CAN FEEL PAIN JUST AS HUMANS DO?

WANT TO LIVE FREE FROM SUFFERING AND HARM?

HAVE THE RIGHT TO BE TREATED WITH CARE AND COMPASSION?

If you answered **YES** to these questions, being an advocate for animals might be your passion as well. Here are some other things that animal activists believe:

ANIMALS ARE OUR FRIENDS, NOT FOOD. Why is it considered okay to eat some animals, like chickens, fish and cows, but not others? In the eyes of animal activists, ALL lives are valued.

LOVE ANIMALS, DON'T WEAR THEM. Cows, sheep, rabbits, foxes, geese and other animals are often used to make clothing and fashion accessories. Activists say that's cruel and inhumane.

ANIMALS DON'T BELONG IN CAGES. Many people visit circuses, zoos and marine parks because they love animals or want to learn more about them. But as activists see it, there's nothing natural or nice about keeping animals captive or forcing them to entertain us.

ANIMAL TESTING IS CRUEL. Millions of animals are harmed or killed each year by companies that use them to test products like cosmetics and shampoo. Millions more are killed so students can dissect them in science classes. Activists think causing innocent animals to suffer is wrong.

PETS SHOULD BE TREATED LIKE FAMILY. For every dog or cat lucky enough to have a good home, countless others are badly treated. Some are abandoned and left to fend for themselves on the streets. Activists stick up for the rights of these animals too.

KIDS TAKING ACTION

"SUBJECTING OTHER LIVING THINGS TO PAIN AND SUFFERING AND DEATH IS THE BIGGEST FAULT IN THE HUMAN RACE."

ZOE ROSENBERG

Born: 2002 **From:** USA

Known for: founding an animal sanctuary

After watching a video about egg production, Zoe gave up eggs and cow's milk. Then she got her parents' permission to turn their backyard into a haven for rescued hens. Zoe has since saved more than 600 animals from factory farms and other bad situations at what became the Happy Hen Animal Sanctuary.

She's also willing to do some pretty crazy stunts to raise awareness about how the food industry treats animals. Like the time she and other activists made a fuss about the hot dogs sold at LA Dodgers' games and ran onto the baseball field with protest banners! Zoe was named Youth Activist of the Year in 2017 and dreams of a day when all animals will be treated with compassion and respect.

SHON GRIFFIN

"BE KIND TO CATS!"

Born: 2012 **From:** USA

Known for: caring for homeless cats

Decked out in his superhero cape, Shon (a.k.a. Catman) is a familiar sight on the streets of his neighbourhood. Since the age of three, he has joined his aunts every weekend to care for stray cats. He goes around filling up cat food bowls and water dishes and giving out chin scratches to kitties that don't have anyone to care for them.

Most of these feral felines are super shy and scamper off if anyone even looks at them. But for some reason they aren't scared of Shon. Not even the biggest scaredy cat of all, a kitty called Bug. Bug had never let any human get near him until the day he first saw Shon. He ran right over and started twirling around his legs! With his new buddy at his side, Bug was checked out by a vet and vaccinated. Now he's one of the friendliest cats in his colony. *Purr-fect!*

CHARLIE SONNEMANN

"TO BE A FULL-BLOWN ACTIVIST AT A YOUNG AGE YOU REALLY NEED AT LEAST A PARENT ON BOARD."

Born: 2004 **From:** Australia

Known for: being an advocate for farm animals

Charlie grew up around his family's backyard 'chooks' – that's Aussie slang for chickens! He loved those chickens and took great pride in caring for them. Then came the day when he realized that his feathered friends were the same birds people ate for dinner. He felt so terrible that when he was twelve, he decided to go vegan. This had a huge impact on those around him. He even inspired his mum, his twin brother and his sister to go vegan too.

Today, Charlie can often be found at vigils and protests or talking to people on street corners about the sad reality of factory farming. He even participated in a 'Fast for Freedom' to protest conditions at one of Australia's largest pig slaughterhouses. Going without food for five days wasn't easy, but it definitely fuelled Charlie's passion for animal rights!

STORM BURGESS

"HUMANS CAN TELL YOU WHEN SOMETHING'S WRONG OR IF THEY ARE BEING NEGLECTED, BUT ANIMALS CAN'T."

Born: 2005 **From:** England

Known for: fundraising for animal charities

When Storm started having some serious health issues, she wanted to find a way to take her mind off her troubles. She decided to make life better for animals in need by raising money for animal organizations. She's such a whiz at arranging events for charity that she's raised an astonishing £30,000 so far – and she's still going! One of her proudest moments was collecting enough cash to give an elderly shelter cat life-saving surgery.

Storm has been honoured at the House of Lords and walked the red carpet at the Animal Hero Awards in London. She also looks after four cats, two dogs, a gerbil and a colony of stray cats. She aspires to be a vet someday. No surprise there!

TAKE ACTION FOR ANIMALS!

FEELING INSPIRED? HERE ARE SIXTEEN THINGS YOU CAN DO TO IMPROVE THE LIVES OF YOUR FURRY FRIENDS:

TALK TO YOUR FAMILY ABOUT EATING MORE PLANT-BASED FOODS. It takes a lot of research and planning to get all the nutrition a growing kid needs on a purely vegan diet (which avoids all animal products, including cheese and milk). Experiment with having meat-free meals once or twice a week and take it from there. More and more schools now offer vegan lunch options too!

HOLD YOUR HORSES!

Not everyone is on the same page when it comes to animal rights. Part of being an activist is listening to other opinions and being respectful, even if you disagree.

RESCUE A PET. Give a loving home to an abandoned animal that longs for a family. Adopting a dog or cat is much better than buying one from a breeder or pet shop.

USE 'CRUELTY-FREE' PRODUCTS. Check the package to make sure it wasn't tested on animals. Look for the leaping bunny logo, a symbol found on many cruelty-free items.

STEER CLEAR OF ANY PLACE THAT MISTREATS ANIMALS. More and more circuses have stopped using animals and that's awesome. Ditto for zoos that do conservation work. Do some research before you visit and be picky about which organizations you support!

DID YOU KNOW?

Many animal activists prefer to say ***pet guardian*** or ***human companion*** rather than ***pet owner.*** Why? They don't see animals as anyone's property!

VOLUNTEER AT A SHELTER. If you are old enough and have your parents' permission, you can help out by walking dogs, cleaning out water bowls, making 'Adopt me!' signs and so on.

RAISE FUNDS for a local animal shelter or animal charity.

SIGN PETITIONS to ban animal testing and other inhumane activities.

WRITE AN ARTICLE for your school newspaper about an animal-related issue.

ASK YOUR SCHOOL TO ADD VEGAN AND VEGETARIAN OPTIONS to the cafeteria menu.

PUT AN ID TAG ON YOUR DOG and ask your parents to have them microchipped too.

KEEP PET CATS SAFE AND HAPPY. Indoor cats enjoy 'enrichment' activities such as cat trees, toys and food puzzles. Any cat allowed to roam outside should wear a collar with an ID tag and also have a microchip in case they get lost.

BE KIND TO STRAY CATS. If your parents and neighbours are okay with it, put out food and water for hungry kitties and/or give them a place to take shelter during the cold winter months.

FOSTER A PET. Sometimes a dog or cat needs a safe place to stay until they are ready to be adopted. You can help loveable pets on their journey to a forever home.

ORGANIZE A PET FOOD AND BLANKET COLLECTION FOR A SHELTER. Check the shelter's website to see what they need. Old towels, cat litter, food bowls and beds are popular items, too!

SUGGEST AN ALTERNATIVE. If your class is planning to dissect an animal, ask for a more humane assignment, such as using a computer program to do a virtual dissection.

SPAY OR NEUTER YOUR PET and encourage others to do the same!

UNIQUE BEGINS WITH 'U'

There's no one right way to live or to be. Yet many of the issues we face today come from a lack of acceptance of others. Diversity activists want to celebrate all the wonderful things that make each one of us special.

Imagine if everyone looked the same, thought the same and acted the same. What if we all had the same talents, the same colour hair and skin, the same likes and dislikes? You know what would happen? We would all be bored out of our skulls, that's what.

This isn't to say that being unique is always easy, especially for kids. It can be hard to put yourself out there. But what's the alternative? **You can only be yourself. Everyone else is taken!**

People are often scared of failing or looking stupid in front of others. But does it really matter what anyone else thinks? Those who can be fearless and embrace everything that makes them one of a kind … those are the ones who create magic with their lives.

Teen activist Greta Thunberg has Asperger's, which is a type of autism. She believes her Asperger's allows her to see the world from a unique point of view. She even calls it her superpower. In her own words, "Autism makes me different and being different is a gift."

SO, IF YOU FEEL DIFFERENT, BE LIKE GRETA AND TREAT IT LIKE A STRENGTH, NOT A FLAW!

Part of being an activist kid is figuring out who you are and what you want to accomplish. So do the world a favour. **Be the best you that you can possibly be.** Speak your truth and follow your wild heart. What is it that makes you **YOU?** Whatever it is, don't hide it. Showcase it!

THERE'S NO ONE LIKE YOU. AND THAT'S YOUR SUPERPOWER.

KIDS TAKING ACTION

BILJANA STOJKOVIĆ

Born: 2005 **From:** Serbia

Known for: inspiring others with her upbeat attitude

"EVERY SINGLE ONE OF US HAS TO LIVE THEIR DREAM AND THEN EVERYTHING IS POSSIBLE."

Biljana's popular video blog – 'I Create My Life', or 'ICML' for short – has made her one of the best-known teenagers in Serbia. She posts on all sorts of topics, but one of her favourites is improving self-esteem – something she knows a lot about. She had bone cancer when she was little but the medical treatment didn't help. By the time she was nine, she was in so much pain that she begged doctors to remove her lower left leg.

Today, Biljana serves as a UNICEF Youth Advocate for Serbia. She uses social media to inspire others and teach children that they have the right to make choices and to change things. As she sees it, the only things we can control in life are our attitude and our efforts. So true!

KHERIS ROGERS

Born: 2006 **From:** USA

Known for: the Flexin' in My Complexion clothing line

"IT DOESN'T MATTER WHAT OTHER PEOPLE THINK ABOUT YOU. IT'S WHAT YOU THINK ABOUT YOURSELF."

Kheris was bullied at school because of her dark brown skin. But she wasn't about to let mean people get her down. She dealt with the playground taunts by telling herself that she was 'flexin' in my complexion', something her grandmother used to say. Kheris liked the phrase so much that she silk-screened it on a T-shirt. That jump-started an entire clothing line designed to inspire people of colour to love their skin.

Kheris went on to make history at age eleven when she became the youngest fashion designer ever to show at New York Fashion Week. She appeared on *America's Next Top*

Model and in an ad campaign for Nike. Kheris also hosts an online show on SoulPancake, *Beyond Bullied*, where she encourages other kids to stand up and stay strong.

HANNAH ALPER

Born: 2003 **From:** Canada

Known for: her motivational speeches and blog

"THE ABILITY TO CREATE CHANGE DOES NOT DISCRIMINATE; IT'S IN EVERYONE."

Hannah launched her popular 'Call Me Hannah' blog at age nine to share her concerns about the environment. Today, she gives TED Talks and speeches all over the place, inspiring people to find their passion and take action for a better world. Hannah isn't one to sit still for very long – she's travelled to Kenya with the charity ME to WE, Costa Rica with National Geographic, and to New York City and Los Angeles for *Teen Vogue* Summits.

She has nearly 40,000 followers on Twitter and serves as an ambassador for Free the Children too. Somewhere in there she also managed to find time to interview other activists for her book, *Momentus: Small Acts, Big Change*. It's based on her belief that small actions can lead to big changes. Busy girl!

NIKKI LILLY CHRISTOU

Born: 2004 **From:** England

Known for: vlogging on YouTube

"LIFE ISN'T ALWAYS UNICORNS AND RAINBOWS. TO GET TO THE RAINBOW, YOU HAVE TO GO THROUGH THE RAIN."

Nikki has a painful condition that causes the right side of her face to swell. She was diagnosed at a young age and soon became quite housebound, so she started vlogging when she was eight. It was her way of dealing with her situation and bringing the outside in.

Nikki does all her own filming and editing, recording her everyday life in a very matter-of-fact way. Putting herself out there in front of millions is an incredibly brave thing to do. Her vlog now has nearly a million subscribers and helps others in ways she never imagined. Nikki received an Emmy Award for a documentary about her life called *My Life: Born to Vlog*. She also won the 2017 *Junior Bake Off* and BAFTA's Special Award in 2019!

BULLY -FREE ZONE

Not everyone out there can see the beauty in things that are different. Take bullies, for example. Many of the kids in this book know what it's like to be bullied. Yet each of them managed to turn a negative into a positive. **Being bullied gave them the motivation to DO something.** Something great.

WHAT IS BULLYING?

Bullying can take many forms, including:

- **name-calling and teasing**
- **spreading rumours**
- **nasty remarks (IRL or online)**
- **being left out**
- **pushing you around**
- **damaging your belongings**
- **physical abuse and threats**

Swedish activist Greta Thunberg was bullied. She even told her mum that she didn't want friends because all children were mean. Now there are *grown-up* bullies saying mean stuff about her – including some world leaders. Greta just shrugs and says they are trying to avoid the real issues at hand. Her critics should be talking about the climate, not her. Duh!

If you are being bullied, it doesn't give you the right to be a bully back. A better thing to do is to keep a diary about anything mean being said or done to you. Jot down names, dates, times and places. Were there any witnesses? Put their names down too. And then (deep breath), show your journal to an adult you trust. And if anyone **EVER** physically threatens you in any way, walk away and tell a teacher or parent **ASAP**.

Bullying is **NOT** okay. Ever. That's why we all need to speak out, so it is stamped out.

"WHEN THE ONLY THING PEOPLE CAN DO IS MOCK YOU . . . IT MEANS THEY HAVE NO ARGUMENT OR NOTHING ELSE TO SAY."

GRETA THUNBERG

ANTI-PLASTIC PIONEERS

Plastic is the enemy. Most of us are well aware of that by now. But how can we avoid the stuff and what can we use instead? We are talking about things beyond the obvious – like saying no to plastic straws.

The war on plastic is one of the biggest challenges the world faces today. There are tons of plastic polluting our oceans and natural habitats and doing serious damage to wildlife. **We've created a throwaway lifestyle, but with plastic, there is no such thing as 'away'.**

Plastic never fully decomposes and a lot of it can't be recycled, either. Pretty much every piece of plastic ever made – more than eight billion tons of it – is still on the planet, releasing toxins and being eaten by animals.

BE A PART OF THE SOLUTION, NOT PART OF THE POLLUTION.

HERE ARE SOME SCARY STATS:

- By 2050 there could be more plastic in our oceans than fish
- 550 million plastic straws are used every day
- 50 per cent of all plastic is used once and thrown out
- Up to one trillion plastic bags are used each day (nearly two million every minute)
- More than 60 million plastic bottles end up in landfills or incinerators every day
- Only 5 per cent of plastic produced is recycled
- Scientists estimate that 8.8 million tons of plastic enter the ocean every year

Plastic pollution is such a huge issue that many environmental activists focus on nothing else. This is one area where they feel they can actually make a difference. **YOU** can make a difference too, each and every day. It's amazing how much plastic we can keep from entering the natural environment just by making a few lifestyle changes!

HOW LONG DOES IT TAKE TO BREAK DOWN?

- **orange peel:** up to two years
- **cigarette butts:** up to ten years
- **plastic bags:** 400 to a thousand years
- **tin cans:** 50 years
- **aluminium cans:** 80 to 200 years
- **plastic bottles:** can last forever
- **glass:** does not degrade (but is much easier to recycle than plastic and is much less damaging to the environment)

ALL THAT GLITTERS

Plastic is often in places you'd never suspect, like glitter. Glitter is so small that it can't be recycled and ends up in rivers and oceans. Skip it or use biodegradable glitter, which takes two to three months to decompose. Still a long time but it's better than forever!

KIDS TAKING ACTION

MELATI AND ISABEL WIJSEN

Born: 2003 and 2001 **From:** Bali
Known for: co-founding Bye Bye Plastic Bags

"DON'T EVER LET ANYONE TELL YOU THAT YOU'RE TOO YOUNG OR YOU WON'T UNDERSTAND."

After seeing the effects of plastic pollution on their Indonesian island's natural beauty, sisters Melati and Isabel decided to take action. Aged ten and twelve they started their own campaign, Bye Bye Plastic Bags, to encourage people to protect the environment. Their tireless efforts have inspired youth groups in twenty-eight countries.

Thanks to their determination and hard work, the entire island of Bali has banned single-use plastic bags, straws and polystyrene. Melati and Isabel are as active as ever, doing interviews, visiting schools, giving TED Talks and speaking out at political meetings.

LILLY PLATT

Born: 2009 **From:** The Netherlands
Known for: founding Lilly's Plastic Pick Up

"EVERY TIME YOU CLEAN UP ONE PIECE OF PLASTIC, YOU SAVE AN ANIMAL'S LIFE. A LOT OF PEOPLE DON'T THINK ABOUT THAT."

When Lilly moved to the Netherlands from London, her grandfather taught her to count in Dutch in a rather unusual way – he took her litter picking. She was stunned to discover that most litter eventually ends up in the ocean. She took a photo of all the rubbish they'd gathered (and counted) and posted it on Facebook. The photo went viral! This inspired Lilly to do more research on things such as the 'microplastic soup' in our oceans.

Lilly is now a school striker and youth ambassador for the Plastic Pollution Coalition. She's met Dr Jane Goodall and world leaders and was featured on TV and in *National Geographic Kids*. To celebrate her tenth birthday she created #LillysGlobalCleanUpDay, a campaign to encourage people to spend a day making the planet a cleaner place.

RALYN 'LILLY' SATIDTANASARN

Born: 2007 **From:** Thailand

Known for: getting stores to ban plastic bags

"I AM A KID AT WAR. I TRY TO STAY OPTIMISTIC BUT I AM ALSO ANGRY. OUR WORLD IS DISAPPEARING."

Lilly became an activist aged eight while at the seaside. She was horrified to see a beach covered in litter. She cleaned it up with her parents, only to wake up the next day to find more rubbish washed ashore. Lilly made it her mission to clean up Thailand, one of the biggest contributors to ocean pollution. She was a small girl willing to take on a gigantic challenge – in her country, the average person uses eight plastic bags every single day!

Inspired by Greta, Lilly did sit-ins in front of Thai government buildings. She skipped school to pull rubbish out of a dirty Bangkok canal. Eventually, people took notice of her efforts and she won her first victory, convincing a major Bangkok supermarket to stop giving out plastic bags once a week. Not long after, more stores pledged to do the same.

THOMAS TRUBY

Born: 2012 **From:** Scotland

Known for: founding the Rubbish Club

"I KEEP DOING IT BECAUSE I JUST WANT THE WORLD TO BE BETTER FOR TURTLES AND OTHER ANIMALS."

Thomas fell in love with sea turtles aged four when he visited an aquarium for the first time. He learned more about them from watching *Blue Planet II*, including the fact that sea turtles often eat plastic bags by mistake because they look like floating jellyfish. This made Thomas so sad that it motivated him to do something. One day when he was out walking his dogs, he started picking up litter along the way. It made him angry that people just dumped their litter. That's when inspiration struck. He'd start a club for litter pickers!

His club now organizes litter picks and beach clean-ups throughout his local region. The Rubbish Club has collected a small mountain of litter so far, as well as things like old tyres, traffic cones, pipes and rusty bikes. Thomas's mum manages his Facebook page, and his posts have inspired people far beyond Scotland's shores.

WHAT YOU CAN DO TO FIGHT PLASTIC!

Think before you buy. Choose a 'naked' apple over one that's wrapped in plastic. Opt for magazines without plastic wrappers and without cheap plastic toys attached. Buy cosmetics that don't contain microbeads. We can all make dozens of choices like this every day! Aim to do what's best for the planet.

Wash clothes carefully. Did you know that many clothes contain plastic? When washed, tiny fibres end up in our waterways, which are then swallowed by fish and other animals. Ask your parents to wash synthetic materials in a specially designed bag or ball that stops the microfibres from getting into the water.

Switch your snacks. Crisp packets and sweet wrappers can take centuries to break down. Opt for homemade biscuits, tins of sweets or things that come wrapped in paper instead. Popping corn can be purchased in bulk. Roasted chickpeas are a great swap for crisps – find recipes online.

Make a plastic-free packed lunch. Use a reusable box, avoid plastic cutlery and use waxed wraps instead of cling film.

Go 'plogging'. This combines jogging with picking up litter. It's fun and keeps you fit too!

Use a reusable water bottle and proper mugs instead of bottled water and paper cups that are lined with plastic and can't be recycled.

Join a beach clean-up. Seven items make up a third of all the litter found on beaches: plastic bags, bottles, lids, cutlery and straws, along with wet wipes, sweet wrappers and crisp packets.

Use flannels instead of wet wipes. Wet wipes are handy but often end up in a sewer fatberg, a congealed mass of non-biodegradable stuff. Ugh!

Remember the bag! Always take a reusable cotton bag whenever you go shopping, and encourage others to do the same.

Stop relying on recycling. Not everything can be recycled into something else. Low-grade stuff ends up in landfills or is burned.

Use a bar of soap instead of shower gel in plastic bottles. Shampoo and conditioner now come in bars too!

Drink bottled milk. Many families are absolutely shocked to learn how many plastic bottles of milk they go through on a weekly basis. The solution is simple – ask your parents to see if milk can be delivered in glass bottles.

Pick it up. Don't just walk past litter – pick it up! This action alone could save a wild animal's life.

EVEN IF YOU DO ONLY MAKE A FEW OF THESE SIMPLE SWITCHES, IT WILL HELP MAKE A BIG DIFFERENCE. SPREAD THE WORD!

Bust the balloons. Balloon releases are really bad for the environment. They eventually end up in the ocean, entangling animals or being mistaken for food. Hang up paper pom-poms for decor instead.

Make your own condiments, such as mustard, chocolate sauce or mayonnaise. Easy to do and you can store them in glass jars.

Sip plastic-free tea. Most people have no clue that there's plastic in their teabags. Loose leaf tea is a much better option and it's easy enough to make.

Use compostable poo bags whenever you're picking up after your furry friend.

Stop chewing plastic. Did you know chewing gum is made from plastic? Ewww! Some companies make natural gum that is biodegradable, plastic-free and chemical-free. It is made with chicle, which comes from a tree.

Get a bamboo toothbrush. Fish don't need to brush their teeth! Keep plastic toothbrushes out of the ocean by quitting them completely. Try silk dental floss too! Look for it online or in health food shops.

Bring your own container for takeaways and skip the Styrofoam (eco-nightmare) and cardboard food containers (most of which are coated with plastic so can't be recycled).

E IS FOR EMPOW

Schools don't always teach about climate change – although they really need to. That means you may need to educate *yourself* about what's going on. You'll make better decisions about your own future that way and ensure your message is loud and clear.

Be the best activist you can be with these simple tips:

Know your stuff. Keep up to date on the issues. Having a few key facts up your sleeve can make all the difference.

Be bold. Don't stay silent because you don't want to make a fuss. It might feel scary the first time you stick your neck out, but you'll be glad you did later.

Stay positive. Learning about all the challenges facing our planet can be quite upsetting. Channel those emotions into positive action! You'll feel better knowing you're trying to change things.

Be proactive. Don't wait to be told to take action on the issues that matter to you. Start your own group. Send off a letter or email. Meet with the head honchos at your school. Keep your eyes and ears open for new opportunities.

ERMENT

BEING EMPOWERED
means you are strong and confident
and in control of your own life.

Prepare for pushback. Not everyone will be on board with your cause or your message. Dealing with negativity is part of being an activist.

Stories beat stats. Share a sixty-second story about what inspired you to become an activist. People LOVE stuff like this!

Walk your talk. Inspire others with your actions and dedication to working for a better world.

Never give up. Activists face challenges all the time. Maybe your family doesn't support you or your friends can't be bothered to stop using plastic. Maybe you're tired or angry or lonely. The best advice is to just ride these feelings out. You'll get through it!

You are part of a movement. Meeting others who share your goals can help you stay motivated and focused. Look for local groups or meet-ups to find other activists near you.

YOU'RE WELCOME!

KIDS TAKING ACTION

YOLA MGOGWANA

Born: 2008 **From:** South Africa

Known for: teaching kids to be eco-warriors

"WE ARE SICK OF WAITING FOR THE ADULTS. WE ARE THE FUTURE AND WE HAVE THE VOICE AND POWER TO CHANGE THINGS."

Yola sees the effects of climate change first-hand. The weather where she lives is hot and dry one day, pouring with rain the next. Not long ago, South Africa had the worst drought in a hundred years and everyone in Cape Town had to ration water. For Yola, all these things are signs that people need to change their ways ASAP.

Yola joined a student strike for the climate and became a volunteer for the Earthchild Project. She helps local schools set up 'living classrooms' – eco-clubs that teach kids about the environment. Yola also shows younger students how to monitor their water use and grow vegetables, even how to reduce food waste by composting with worm farms!

ZURIEL ODUWOLE

Born: 2002 **From:** USA

Known for: using film to promote girls' education

"I AM CONCERNED THAT NOT EVERY GIRL HAS A RIGHT TO GET AN EDUCATION OR A CHANCE TO ACCOMPLISH HER DREAMS."

When Zuriel was nine, she entered a documentary-making contest at her school. She wanted to do something different, so she went to Africa to interview Ghana's president. She noticed that many girls her age weren't in school there and decided to start an education programme called 'Dream Up, Speak Up, Stand Up' for girls in Africa.

Zuriel has interviewed political leaders and made other documentaries about Africa. She's also been interviewed by the BBC and at ten became the youngest person featured in *Forbes* magazine. Listed as one of Africa's '100 Most Influential People', Zuriel now teaches filmmaking to young girls all over Africa.

ALEX WHITE

Born: 2003 **From:** England

Known for: blogging about nature and wildlife

"IF THE BOOK INSPIRES JUST ONE EXTRA PERSON TO BECOME EXCITED ABOUT NATURE AND WILDLIFE THEN I'LL BE HAPPY."

Alex's interest in wildlife began aged three when he first went badger watching with his parents. He started his own blog aged ten to share stories about local wildlife. His blog features many of his animal photos, like his prize-winning shot of a wide-eyed hare.

In his first book, *Get Your Boots On*, Alex offers helpful hints for getting closer to nature, including when and where to go, what to look for and what to bring along. His goal is to inspire other young people to get outside and perform what he calls 'random acts of wildness'. As Alex sees it, there's no need to go long distances to have adventures or see amazing landscapes and animals. All these things can be found right on your doorstep!

NIKHIYA SHAMSHER

Born: 2003 **From:** India

Known for: founding the Bags, Books and Blessings project

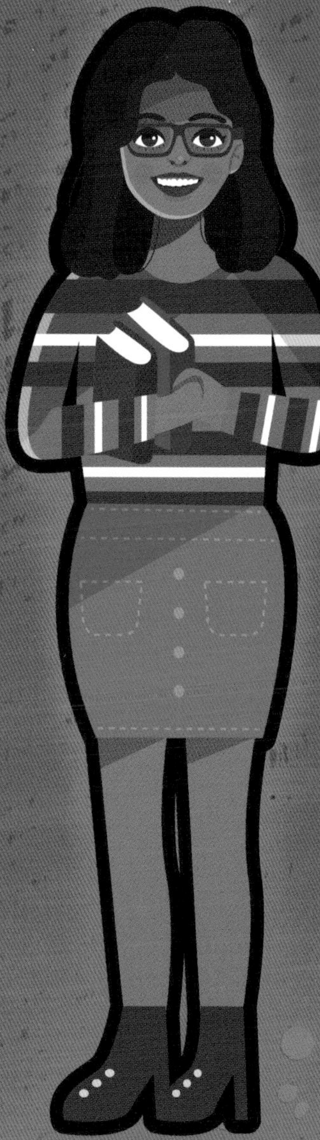

Nikhiya's activist journey began when she casually gave her old school bag to a girl from a poor family. The girl was overjoyed and said it was the first school bag she'd ever had. It gave her confidence and made her excited to go to school. Nikhiya had no idea that such a simple item could mean so much to someone.

Nikhiya was shocked to learn that as many as fifty students in a school might share a single textbook. Many walked barefoot to school, so a book bag was a luxury. That gave Nikhiya the idea for 'Bags, Books and Blessings'. Thanks to her, thousands of kids now receive school supplies, uniforms and shoes, all donated by local businesses. In another project, Nikhiya funds science labs in schools that don't have them. She was recently honoured with the Diana Award.

"I BELIEVE THE BEST WAY TO PREDICT THE FUTURE IS TO CREATE IT."

SCHOOL PROJECTS THAT MAKE A DIFFERENCE!

How earth-friendly is your school? Chances are, there could be some improvements and who better to suggest them than an activist as informed as **YOU**? Once people know the facts, they may be much more empowered to help. And young people are more likely to listen to eco-activists their own age!

01 Build an **igloo** from plastic milk containers or eco-bricks (plastic bottles stuffed with clean, dry plastic bags and used as building blocks).

02 Ask if each classroom can have **bins for recycling AND rubbish**. Sorting stuff makes recycling much easier!

03 Have a **school 'green team'** to help educate students about climate change. Members could meet each month to discuss new projects.

04 Set up a **battery recycling station**. Everyone has old batteries!

05 Start a **bike-riding campaign** to encourage people to use pedal power instead of cars.

06 Plan a **litter-picking** day to keep the school grounds tidy.

07 Turn plastic milk cartons into **bird feeders** for the school playground.

08 **Grow vegetables!** Lots of schools have their own garden patches for students to tend.

09 **Show a nature or wildlife documentary**, like the *Blue Planet* series or *Drowning in Plastic*.

10 Host an **eco-quiz competition** for the whole school.

MIND THE GAP!

Taking a year out between school and university is becoming a popular option for older teens. Some choose to volunteer for environmental projects, such as protecting coral reefs, standing guard over turtle eggs, cleaning up litter-strewn beaches and much more. Swedish activist Greta Thunberg is taking a gap year to focus on her activism. She started off by crossing the ocean for a UN meeting, travelling aboard an eco-friendly yacht!

11 **Visit a recycling plant**. You'll learn some amazing things and your classmates may be inspired to Reduce, Reuse, Recycle!

12 **Have a junk model contest**. Teams can make models using egg cartons, cardboard tubes, tin cans, cereal boxes and plastic bottles. Robots, vehicles, the Eiffel Tower … anything goes!

13 Host a **fashion show** featuring items made with charity shop finds or upcycled materials.

14 **Start a new club**, such as a Gardening Club, Recycling Club or Eco-Activists Club.

15 **Go on an eco-adventure**. Activities like pond dipping or river walks can be a fun way to discover nature. Take clipboards, magnifying glasses and a camera to record what you find.

NATURAL WONDERS

Like people, animals and plants need a place to live. But humans are chopping down forests and polluting oceans, rivers and other natural habitats at an alarming rate.

This damage is especially bad in places like the Amazon rainforest. Every minute, an area the size of a football pitch is destroyed, so that the land can be used for farming, livestock or houses. At this rate, all rainforests could disappear in the next hundred years! Tragic. And the fires that raged out of control in the Amazon in 2019 only made things worse.

Habitat loss is a big threat to endangered plants and animals. This is a problem for all sorts of reasons, including the fact that all species on earth are interconnected. From tiny algae to tall redwood trees to human beings, everything is carefully balanced and dependent on each other. If one species becomes endangered, the less variety (or 'biodiversity') there is and the weaker these connections become. That's bad news for everything on the planet.

More plant species are threatened with extinction today than at any other time in history. As we are finally realizing, nature is not simply a warehouse of resources that can serve human needs forever. We depend on nature for our very survival and need to treat it with care and respect. If we don't stop the damage we're doing, we risk losing everything.

Experts say that Mother Nature urgently needs our help to recover – but we need to act **NOW!**

"I WANT YOU TO ACT AS IF OUR HOUSE IS ON FIRE. BECAUSE IT IS."
GRETA THUNBERG

That's where you come in. Young activists around the world are joining together to fight for a greener future. They believe the earth is borrowed and that each generation should leave something behind – something better – for the next generation. If you believe that too, keep reading to find out how to help!

KIDS TAKING ACTION

MADISON PEARL EDWARDS

"NEVER UNDERESTIMATE THE POWER OF AN ANGRY CHILD ON THE INTERNET, YELLING AT THE WORLD, 'LEAVE MY DOLPHINS ALONE!'"

Born: 2006 **From:** Belize

Known for: saving a barrier reef

Madison lives in Belize, home to earth's second largest barrier reef. When she heard of plans to drill for oil nearby, she couldn't believe anyone would put such a precious place at risk, along with the wildlife there. She shared her outrage on her blog and social media.

Madison also wrote about snorkelling the length of the barrier reef – it took a week and she was the first Belizean to do it! Madison's efforts inspired others to help by signing petitions, sending emails and putting pressure on the government. Belize has now passed a law that bans oil drilling anywhere in its marine territories. Success!

CATARINA LORENZO

"IF [WORLD LEADERS] DON'T ACT TO STOP THE CLIMATE CRISIS, IT IS OUR FUTURE THAT WILL BE AFFECTED."

Born: 2007 **From:** Brazil

Known for: petitioning the UN on behalf of young people

Catarina spends a lot of time on Brazil's beaches and has felt the rising temperatures in the ocean and seen first-hand the toll it is taking on coral reefs and local marine life. Catarina has also experienced an increasingly drier climate due to lack of rain, which makes forest fires worse and causes water shortages that can last for days.

Storms are also getting much worse. In July 2019, a bad storm blew in with such heavy rain and strong winds that it damaged Catarina's home and dumped toxic sewage into streams and onto beaches. She became so concerned that she was one of sixteen young people to formally complain about climate change to the United Nations.

IRIS DUQUESNE

Born: 2003 **From:** France

Known for: petitioning the UN on behalf of young people

"IT JUST MAKES ME ANGRY. [OUR GOVERNMENT IS] SUPPOSED TO PROTECT US."

The first summer of Iris's life was the hottest in Europe since 1540. She was three months old when a deadly heatwave swept across France, making her parents fear for her life. As she has grown up, southwestern France has continued to experience extreme weather caused by global warming. Deadly floods are common and storms hit more often and with greater intensity. During one storm, hail the size of golf balls rained down on the area! Tides are getting dangerously high too, pushing the coastline closer inland. One of Bordeaux's largest buildings was even evacuated, as people feared it might fall into the ocean. Iris is stressed out about climate change and feels powerless and betrayed, so much so that she finally joined other young activists to deliver a formal protest to the UN.

DARA MCANULTY

Born: 2004 **From:** Northern Ireland

Known for: campaigning for equal access to nature

"ONE BIG POSITIVE ABOUT HAVING AUTISM? WE NEVER GIVE UP, WE KEEP GOING UNTIL THE JOB IS DONE."

Dara devotes all his spare time to helping nature and wildlife, but it is far more than just a hobby. He has Asperger's and being in nature makes him feel more connected to a world that can often seem confusing and overwhelming.

When Dara's family moved from Belfast to the countryside, he started a blog to share his love of nature. He also produces videos, gives talks and does beach clean-ups. He once did a thirty-mile mountain trek (in the winter!) to raise money to prevent wildlife crime.

Dara is the youngest recipient ever of the RSPB Medal, awarded by the Royal Society for the Protection of Birds. He has also been featured on *Blue Peter*, the British kids' TV programme. Somehow he still found time to author a book, *Diary of a Young Naturalist.*

LIVING A FRILUFTSLIV LIFE

What's *friluftsliv*? It's a Swedish word that refers to a way of life that is spent exploring and appreciating nature. The activity isn't important – it's more about taking every opportunity you have to spend time outdoors. So what are you waiting for? Go outside and play!

TAKE A WALK around your neighbourhood or a nearby park.

PACK A PICNIC. Food seems to taste better when it's eaten outdoors!

ATTRACT BEES. Vegetables and flowers are fun to grow and help to attract pollinators. Even if you don't have much room, you can plant some herbs in small pots on a windowsill. You might also leave part of the garden 'wild' so that meadow flowers and grasses can flourish.

SLEEP UNDER THE STARS. Spend the night in a tent in your back garden or camp on the moors, in a forest, near a beach, in the mountains, wherever!

CREATE LANDSCAPE ART with twigs or fallen leaves or stones. Then take a photograph and leave it all behind.

CLIMB A HILL. Race your friends to see who can reach the top first!

PICK APPLES at an orchard and turn them into something yummy, like a fruit pie or crumble.

SEARCH FOR CONKERS, then challenge your friends to a game!

GO FOR A HIKE with your family or join your local Scouts or Guides for lots of outdoor adventures.

PLAY A GAME OF NATURE I SPY on the way to school.

SMELL THE AIR. How many different scents can you detect? Flowers? Rain? Grass? Earth?

MORE GREAT WORDS TO KNOW

Gökotta is a Swedish word that means 'to rise at dawn to listen to the birds sing'.

Earthing is walking barefoot to directly connect with the energy of the planet. Some studies say it's good for your health.

Shinrin-yoku is a Japanese expression that means 'forest bathing'. Find a place with lots of trees and sit still or walk in silence. Just BE.

GAZE AT THE MOON. Is it full, new, waxing or waning? Knowing the different lunar phases makes moon viewing even more fun.

SIGN UP FOR A NATURE ACTIVITY. There's something out there for everyone, including hiking, canoeing, outdoor survival skills, orienteering, astronomy, forest skills and nature photography.

CREATE A WILD INSTAGRAM. Take a photo of a plant or landscape every day for a month and watch your Instagram grid come alive with colour and life. (Remember – you should be at least thirteen to have an Instagram account!)

GO OUT AFTER DARK. Join a badger walk or bat watch with an expert guide to meet some wonderful night-time wanderers.

GRAB YOUR WELLIES. So what if it's rainy and gloomy? There's a famous Swedish saying: "There's no such thing as bad weather, only bad clothes." Put on your boots and raincoat and head out anyway! It's amazing how different things can look, sound and smell in the rain.

HOW TO BE A WILD CHILD

Your actions now can help preserve the natural world for generations to come. Here are some easy ways to get started:

Avoid touching or picking plants growing in the wild. Studies show that plants don't really like to be handled. Be sure to leave them where they are so others can enjoy them too.

Steer clear of pesticides, which can get into water systems and harm plants and wildlife (and pets and people!).

Visit national parks and nature reserves. Such places often rely on tourism to pay for conservation work. Try to go off-season, when there are fewer visitors.

Raise money for your favourite nature organization with a cake sale, car boot sale or by selling handmade crafts.

Stick to the trail to avoid damaging vegetation or the animals that live there.

Join an eco-club or conservation programme for kids.

Learn to identify plants using books or a smart phone app.

DID YOU KNOW?

- Bamboo is nature's fastest-growing plant. Some species can grow 91 centimetres a day.
- About 15 billion trees are chopped down every year just for paper.
- Tropical rainforests are home to 80 per cent of all species.

Raise awareness about nature conservation by creating a website or blog.

Save paper. Cut down on waste by printing only when necessary and on both sides. Write notes on old envelopes and send e-cards. Fill up each sheet as much as possible before recycling it. Maybe your teacher can switch from printed homework to online assignments?

Use reef-friendly sunscreen that won't cause coral bleaching or harm marine wildlife.

Be choosy with your snacks, many of which are made with palm oil. Palm oil is BAD. All over the world, rainforests are being destroyed to make way for huge palm oil plantations. Read labels to make sure the biscuits and crisps you eat don't contain the stuff and boycott the ones that do.

Recycle or repair your electronics to keep toxic chemicals from polluting soil and water near landfills. Some charities (such as Oxfam) even offer phone-recycling services!

Feed the birds. Hang a simple feeder in your garden or attach one to a window.

Leave things alone. Let ivy climb up walls to create habitat and food for insects. Give slugs a chance to eat plants in the garden. Think of weeds as wildflowers and suddenly they look a lot prettier!

INGENIOUS IMAGINEERS

Many people think that young people are glued to their phones and obsessed with technology, but plenty of them are also improving the world with their clever inventions. They are natural solution seekers with incredible imaginations and a great idea is a great idea, no matter who comes up with it or how old they are!

Here are just a few things invented by young people throughout history:

ADDING MACHINE. Blaise Pascal created an early version of a calculator in 1642 aged eighteen.

BRAILLE. When Louis Braille was twelve, he came up with an easier way for him and other blind people to read. His system uses a series of raised dots that readers feel with their fingertips. It can also be used to write.

EARMUFFS. Fifteen-year-old Chester Greenwood wanted to keep his ears warm when he went ice skating. He asked his granny to sew pieces of fur on the ends of two strands of wire. So many people wanted a similar pair that he got a patent and marketed his invention as ear mufflers.

GLOW-IN-THE-DARK PAPER. Twelve-year-old Becky Schroeder created glow-in-the-dark paper in 1974, an invention that became incredibly useful for doctors and photographers who needed to write in the dark.

HUG MACHINE. Temple Grandin was eighteen when she invented a deep-pressure device used to relieve stress. Temple had autism and came up with the idea in the 1960s after she observed how a similar contraption helped calm frightened farm animals.

"THE TRUE SIGN OF INTELLIGENCE IS NOT KNOWLEDGE BUT IMAGINATION."

ALBERT EINSTEIN

ICE LOLLY. Frank Epperson accidentally invented this sweet treat in 1905, when he was eleven. After making a drink by stirring together water and flavoured powder, he left it on the front porch overnight with the stirring stick still inside. He found it frozen the next morning. Eureka!

LOOM SAFETY DEVICE. Margaret Knight worked at a textile mill in the mid-1800s. She was twelve when she witnessed a terrible accident that inspired her to create a safety device for textile looms. She went on to invent many other things, including the flat-bottomed paper bag!

TELEPHONE. Alexander Graham Bell was eighteen when he figured out how to use an electrical current to send sounds from one place to another. His experiments led to the telephone.

TELEVISION. Philo Farnsworth was fourteen when he sketched out an idea for a device that could display electronically produced images. His prototype led to the world's first telly in 1927.

TRAMPOLINE. When George Nissen was sixteen years old, he invented the trampoline to help him gain muscle mass for gymnastics. It was an immediate hit in the mid-1930s.

How's that for some inspiration? Read on to learn more about four young inventors putting their imaginations to good use!

KIDS TAKING ACTION

XÓCHITL GUADALUPE CRUZ LÓPEZ

"THE TRUTH IS THAT SCIENCE IS SOMETHING VERY BEAUTIFUL AND EXCITING."

Born: 2009 **From:** Mexico

Known for: inventing a solar-powered water heater

Xóchitl lives in a very cold rural region of Mexico. Locals heat water with wood fires instead of more costly electricity. Xóchitl decided to find a cheaper, earth-friendly solution.

Starting with sketches, Xóchitl then gathered recycled materials to build her idea: 15 metres of black hose, ten sturdy plastic bottles painted black, plastic cable ties, a wooden base, black nylon and recycled glass. With her dad's help, her solar heater was installed on the roof of their house. It worked! It could heat 10 litres of water to 35 to 45°C.

For her invention, Xóchitl became the first child to ever win the ICN Women's Recognition Award from Mexico's National Autonomous University. She was only eight at the time!

MIHIKA SHARMA

"I HOPE IN THE FUTURE THAT ALL THE INVENTIONS I MAKE CAN COME TRUE AND HELP LOADS OF PEOPLE."

Born: 2010 **From:** England

Known for: inventing the Smart Stick

Mihika was upset when she saw a blind woman stumble while crossing a road. She decided to design a tool to help visually impaired people. The Smart Stick, as she named it, had tons of features: vibrating alerts, LED lights so it could be seen in the dark and a camera to recognize and track obstacles. It even had a sensor to detect puddles!

The idea was so good that students at University College London helped her build a prototype with a 3D printer. It went on to win the 2019 BT Young Pioneer Award. Wow!

"YOU ARE NEVER TOO YOUNG OR TOO OLD OR TOO INEXPERIENCED TO SEE A PROBLEM IN THIS WORLD AND TRY TO SOLVE IT."

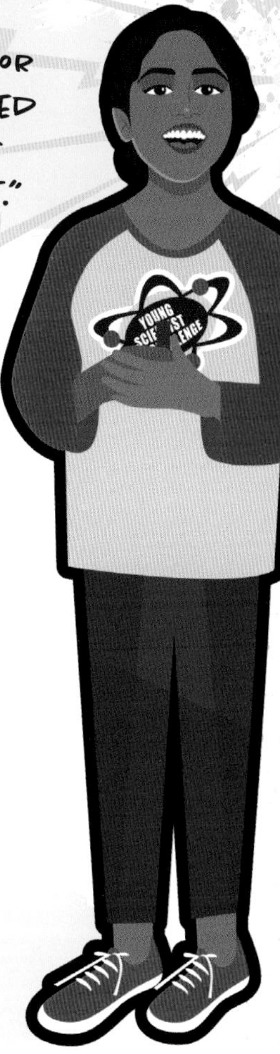

MAANASA MENDU

Born: 2003 **From:** USA

Known for: inventing a clean energy device

On her family's annual visits to India, Maanasa saw with her own eyes what life is like when people don't have electricity or clean water. For 50 million people there, life practically stops at sunset. Maanasa saw children crowd around a single kerosene lamp to do their homework.

Inspired to make a difference, Maanasa designed a device that creates renewable energy from the sun, wind and rain. It uses 'solar leaves' and a special material that generates electrical currents from vibrations. Maanasa named her invention HARVEST because it has the ability to gather energy. Each device is also incredibly cheap to make – less than £5. Maanasa was named America's Top Young Scientist and hopes to introduce her invention worldwide as soon as the design is perfected.

"I'M NOT SAYING MY PROJECT IS THE SOLUTION. THE SOLUTION IS THAT WE STOP USING PLASTIC ALTOGETHER."

FIONN FERREIRA

Born: 2001 **From:** Ireland

Known for: removing microplastics from water

Fionn's light-bulb moment came when he found a rock covered in oil near his small coastal town. He noticed bits of microplastic stuck to the oil. Too tiny to be removed by filters at waste-processing plants, these microplastics end up in waterways, and eventually into wildlife and humans.

The rock reminded Fionn of something he'd learned in chemistry class: 'like attracts like'. What if microplastics had something to stick to? He came up with a solution: a liquid that attracts plastic particles and can be removed with a magnet, leaving clear water behind.

Fionn entered this project in the 2019 Google Science Fair and came away with the $50,000 top prize! He hopes his solution can be used at wastewater treatment facilities to prevent microplastics from reaching the ocean.

CURIOUSER AND CURIOUSER

Being an imagineer means looking at problems as challenges – challenges that can be solved with some creative thinking. Here are some tips to become a more out-of-the-box thinker:

Be positive. Don't assume anything is impossible. Think about why something might work, rather than why it might not.

Get a fresh perspective. Try looking at the challenge from a different angle. How do they do things in other cultures, for example? Keep an open mind and you'll find inspiration everywhere.

Observe. Pay close attention to the world around you and learn from others by observing them. A breakthrough is sometimes right in front of you!

Expand your circle. Spend time with kids from outside your usual circle of friends. Hanging out with different people might give you some new ideas or fresh ways of seeing things.

Try new things. Doing stuff you've never done before helps your brain switch gears and generate new ideas. Try new foods, learn a new skill, sign up for a workshop, whatever!

Don't be afraid of looking silly. Some great ideas may come when you let your guard down and let whatever happens happen.

THE DESIGN PROCESS

IDENTIFY A NEED. First things first: find a problem that needs to be solved. You don't have to look far for stuff that could use some improvement – start with your home, community or school.

BRAINSTORM. Start thinking of possible solutions and do some research. To understand the problem even better, talk to the people most affected by it. You might hear something that gives you a great idea.

DESIGN AND BUILD. Once you've hit upon an idea, start to think about your solution. Sketch it out on paper and make a list of all the materials you will need. Find someone who can help answer questions as you move forward, like a parent or teacher.

TEST AND FINE-TUNE. Run some tests to perfect your invention. Ask other people to test it out too and take notes on their feedback. Keep tinkering and don't get discouraged! This stage is all about trying and trying again.

SHARE YOUR SOLUTION. Let the world know about your invention. One of the best ways to do this is to enter a science fair or competition. Who knows what the future holds for your creation?

Ask 'What if?' Innovators are curious and ask a lot of questions.

Ask 'Why?' Why are things done like this and not that? Just because something has always been done a certain way doesn't mean that's the only way. Or the best way.

Relax. The biggest bursts of creativity tend to happen when you give your brain a break. Run a bath, take a nap, gaze at clouds. Chilling out gives your grey matter time and space to work.

Change up your room. Your 'creative space' is very important. Put away anything that might distract your attention and replace with things that inspire you – maybe a vision board, a poster of a favourite scene or motto, a different colour of wall paint (ask your parents first!). Whatever makes you feel creative.

Keep an ideas journal. Have a notebook or a computer file where you write down random ideas, even half-baked or seemingly impossible ones. Sometimes these are the best ones!

TOO. MUCH. STUFF!

You don't need to be a rocket scientist to know that there's a lot of rubbish out there. Our landfills and recycling centres are overflowing. Many things don't even make it that far. Think of all the litter along roadsides or wildlife with plastic bags in their bellies.

Somewhere along the line we started wasting resources, buying stuff we don't really need … and mindlessly tossing it out when we're done. We've created a throwaway society. But everything we throw away doesn't just vanish into thin air. All that waste keeps GROWING and GROWING and GROWING. It's killing wildlife and destroying our beautiful planet.

Activists around the globe are trying to get people to produce less waste and make better use of what they already have. If you take away just one thing from this chapter, let it be this:

The true measure of success and happiness in life is NOT how much stuff you have.

"WE DON'T NEED A HANDFUL OF PEOPLE DOING ZERO-WASTE PERFECTLY. WE NEED MILLIONS OF PEOPLE DOING IT IMPERFECTLY."

ANNE-MARIE BONNEAU

Some cultures have figured this out already. There's even a word for it in Swedish: *lagom* (pronounced lah-gom). It means 'everything in moderation' or 'not too little or too much'. People there try leaving something behind for others and not taking more than they need.

Another good example of this sort of thinking comes from Marie Kondo. She wrote a very famous book that challenges people to take a hard look at all their stuff. Her advice is to cut down on clutter by keeping only the things that 'spark joy'. It's a simple philosophy that has been life-changing for millions of people.

Your choices make an impact on the planet every single day. So aim to do the right thing and inspire others to do the same!

KIDS TAKING ACTION

RYAN HICKMAN

Born: 2011 **From:** USA

Known for: founding Ryan's Recycling Company

"I THINK IT'S REALLY COOL THAT I CAN MAKE A DIFFERENCE IN THE WORLD BY FOLLOWING MY PASSION FOR RECYCLING."

Ryan discovered his life's passion aged three, when he visited a local recycling centre with his dad. He loved getting money for feeding cans and bottles and glass into the machines and it motivated him to do more. The next day he announced that he had a new business: he was going to pick up all the cans and bottles from everyone in the local area.

His favourite activity became helping to separate the recycling at school. By seven, he was the boss of his own company, which has recycled nearly 500,000 cans and bottles (so far!). Ryan's efforts to keep rubbish from going into landfills and oceans have earned him many awards, as well as thousands of dollars for his future education. He donates the proceeds from T-shirts sold on his website to a local marine life rescue organization.

MADDISON MCQUEEN-DAVIES

Born: 2002 **From:** New Zealand

Known for: founding Share a Pair NZ

"IT'S VERY HUMBLING TO SEE HOW MUCH OF A NEED THERE IS."

Maddison was shocked when her mum said that children at the school she worked at often had worn-out shoes. The end of the school year was approaching and Maddison realized that many students would outgrow their shoes over the break, or no longer need them if they'd left school. Light-bulb moment! And the inspiration for Share a Pair NZ.

Maddison put notices up at her school asking for pre-loved shoes and set up a Facebook page to spread the news. Thanks to Maddison, thousands of shoes and trainers have been donated to kids in need, as well as footwear for refugee families. Maddison hopes to expand the project nationwide. She received the Diana Award for her work in 2019.

STORM DELANEY

Born: 2002 **From:** Australia

Known for: creating eco-chic fashion

"EVERYTHING WE'VE USED, WE CAN MAKE SOMETHING BEAUTIFUL OUT OF IT. IT DOESN'T HAVE TO BE THROWN AWAY."

Teen model and designer, Storm, has a passion for fashion. She's made a name for herself by crafting stunning outfits out of materials destined for landfill. Her up-cycled, experimental designs also deliver a clear message about the environment.

One favourite piece looked like a shimmering silver and black leather outfit from afar but was actually made with one thousand ring pulls held together with fence links! She used twenty more ring pulls to make a choker necklace to wear with it. The entire outfit cost next to nothing but looked like high fashion.

For a nature lover like Storm, designing eco-clothing is her way of protesting 'fast fashion' – poorly made cheap and trendy clothes that take a terrible toll on the planet.

NIKITA SHULGA and SOPHIA-KHRYSTYNA BORYSIUK

Born: 2006 **From:** Ukraine

Known for: co-founding Compola, a composting project

"WE SHOULD SOLVE THESE PROBLEMS, NOT CLOSE OUR EYES . . . BECAUSE IF YOU DON'T DO IT, THEN NOTHING CHANGES."

Nikita and Sophia-Khrystyna were upset by the amount of food being wasted by their school canteen. They knew that scraps could easily be turned into compost – an organic fertilizer that can help create healthy soil. It would be a big step forward for their country, where landfills are overflowing and very little is recycled.

The teens offered to install a composter at their school to recycle organic waste. The school agreed, but only if they found a way to pay for it. With a crowdfunding campaign, they did! They named their project 'Compola' and taught their classmates how to sort rubbish and compost waste. They even started a competition with 200 other schools. Their ultimate goal is for all 18,000 Ukrainian schools to compost their food leftovers.

THE SIX 'Rs' OF RECYCLING

01 **REFUSE.** Ever noticed how people are always trying to give you stuff? A flyer on the street, a sample in a store, a goodie bag at a birthday party? Next time, just say, "No, thanks!"

02 **RE-USE.** Avoid disposables and opt for things that can be used over and over, like stainless-steel water bottles instead of single-use plastic ones.

03 **REDUCE.** How much of your stuff do you actually use or love? Whatever doesn't make the cut needs to move on to be loved by someone else. Donate to a charity or give things to friends or family. Having fewer things also makes your room look instantly tidier (a good way to score points with your parents).

TIP: WHEN IN DOUBT, THROW IT OUT!

If you aren't sure something can be recycled, put it in the rubbish bin. When non-recyclable stuff is mixed in with the rest, it means all of the recyclable items will be dumped too.

04

RECYCLE. Recycling takes things like plastic, paper, metal and glass and breaks them down so they can be remade into a new product. It's better than throwing things away but should be a last resort. Why? Many recyclables still end up in landfills or oceans (literally tons of it). Recycling makes us feel like we are doing something good but striving to create NO waste at all is even better!

05

REPURPOSE. Factories emit CO_2 when making new products, so instead of buying things that are new, fix your old stuff, mend your clothes or have a swap night with friends. A bit of creativity is all it takes to transform you into a waste-fighting, treasure-making superstar!

06

ROT. Let organic material (like food scraps) rot in a compost pile. Putting compostable stuff in with general waste releases harmful methane gas when it decomposes. Composting limits the gas and creates a great fertilizer for plants.

BE A ZERO-WASTE HERO

A 'zero waster' is someone seriously committed to having less stuff in their life. Their goal is to generate as little waste as possible. They wear second-hand clothes, refuse to buy plastic and take their own bags, containers and jars to shops. They use bar shampoo and bar soap, buy food that's about to go off, make things from scratch and compost their leftovers.

Some zero wasters are so committed that they can store an entire year's waste in a single glass jar! Not everyone can do that, and definitely not overnight. For now, aim for a low-waste life rather than a zero-waste life. Focus on making one small change each day and all those changes will start to add up to make a real difference. Here are some ideas:

LET IT GO! Donate stuff you no longer have any use for, or be part of the increasingly popular Generation Share movement and swap or borrow clothes from friends.

GET CRAFTY. Turning trash into treasure is a fantastic creative outlet. There are tons of great ideas online.

SECOND-HAND CHIC. Buy pre-loved clothing from a charity shop or online site rather than something new. It helps save resources and vintage clothing is often better made and more unique.

FORGET 'FAST FASHION'. Cheap clothes aren't bargains if you have to keep buying more.

SPEND YOUR MONEY ON EXPERIENCES, NOT STUFF. It's better to make memories than to create clutter.

NIX THE WOODEN FORKS. Swapping one type of disposable packaging for another is NOT eco-friendly. It's wrong to chop down forests to make items that we use only for a few moments. A better way would be to switch to a reuse and refill model – things like metal utensils, proper teacups and selling things in bulk at refill stations.

GET CREATIVE WITH UP-CYCLING. Up-cycling is the process of turning old or discarded items into something useful and even beautiful. In the words of *Sesame Street*'s Oscar the Grouch, "Just because you're trash doesn't mean you can't do great things. It's called a garbage can, not a garbage cannot."

CREATIVE FOR A CAUSE

There are countless ways to use your creative talents to work as an activist. You can spread a message with a song or photo, or motivate someone to act with a letter or speech. Get people's attention with a piece of art, or share stories through a performance or video you've made.

For any change to happen, people need to feel something. Maybe it's anger or frustration or fear. Maybe it's love or compassion or a sense of hope. **Creative activism** can deliver both.

When people talk about the 'creative arts', they used to mean things like:

FINE ART | DRAMA | LITERATURE | DANCE | MUSIC

These days, there are hundreds (thousands?) of other ways to express your creativity:

BLOGGING | MAKING VIDEOS | INSTAGRAM PHOTOS
STREET ART (LIKE WALL MURALS OR GRAFFITI)
HANDCRAFTED ITEMS | FASHION DESIGN
RAP MUSIC | YARN-BOMBING (GOOGLE IT!)
ANIME | CODING (YES, REALLY) | COOKING
JEWELLERY-MAKING | BAKING

There's more – way more – but there's only so much space on this page!

Being creative is all about expressing yourself or making others think or feel something. And the best part? You don't need to be able to draw or sing or belong to the school drama club. **We are ALL creative, each in our own way.** Being creative is part of what makes us human. We are born with a sense of curiosity and endless imagination. These things help us find solutions, take risks and face our fears.

The arts have been a form of communication for centuries. And whether or not you are 'good' at it, you can still create something. There's no better time than now!

"YOU CAN'T USE UP CREATIVITY. THE MORE YOU USE, THE MORE YOU HAVE."

MAYA ANGELOU

KIDS TAKING ACTION

MARLEY DIAS

Born: 2005 **From:** USA

Known for: #1000BlackGirlBooks campaign

"FRUSTRATION IS FUEL THAT CAN LEAD TO THE DEVELOPMENT OF AN INNOVATIVE AND USEFUL IDEA."

Marley loves books. By the age of eleven, she was going through stacks and stacks of them. But she got fed up with the books assigned at school. They all seemed to be about white boys and their dogs. When Marley searched for books with girls of colour as the main character, she couldn't believe how few there were. What was up with THAT?

She knew if she was frustrated, she wasn't the only one. So she decided to collect 1,000 books about black girls and give them away to other black girls like her. Marley hit that goal and kept on going. She gathered more than 9,000 books! Then she went on to write her own book – *Marley Dias Gets It Done* – to inspire other kids to become activists too.

SOFFIA CORREIA

Born: 2004 **From:** Brazil

Known for: rap music that empowers girls

"I WANTED TO HELP TELL OTHER BLACK KIDS THAT THEY SHOULD BE PROUD OF WHO THEY ARE."

Soffia is a rap artist who goes by the stage name MC Soffia. She started learning how to DJ, rap, breakdance and create street art when she was only six. A few years later, a classmate teased her about her skin colour. She said it was so dark that Soffia must have fallen into a tin of paint. Soffia was so upset that she wrote a song about it.

That song and many others have inspired girls and women of colour to take pride in their skin. One of Soffia's most popular songs is about famous black Brazilian women. Another song, 'Barbie Black', says you don't need to be tall, skinny and blonde to be beautiful. Soffia is now one of her country's biggest hip-hop stars and performed at the opening ceremony of the Rio de Janeiro 2016 Olympic Games.

DAVID WICKER

Born: 2005 **From:** Italy

Known for: coordinating communications for student strikes

"WE WILL NOT STOP UNTIL POLITICIANS AND LEADERS DECIDE TO TAKE ACTION."

Computer programming may not seem like a creative activity, but it definitely is. It's all about solving problems – something that requires a different way of thinking. That's David's contribution to the #FridaysForFuture movement.

David noticed that student strike organizers were getting the word out but using multiple platforms to do it. There were posts on Messenger, WhatsApp, Facebook and other social media channels. David thought it would be better if everyone used just one site.

Connecting and transferring everything to a single platform keeps David busy, but he still attends weekly school strikes. More young people than ever are striking in Italy after the education minister urged schools to let students take part in climate change protests. Bravo!

RAHUL AND ROHAN RAJU

Born: 2003 and 2005 **From:** United Arab Emirates (UAE)

Known for: artwork featuring endangered species

"AN ELEPHANT IS KILLED EVERY FIFTEEN MINUTES ... WE DON'T HAVE MANY OF THESE BEAUTIFUL ANIMALS LEFT."

Rahul and Rohan's Save Wildlife campaign was inspired by the plight of endangered species. They want to use their art to save wild animals by raising awareness of poaching.

The brothers started painting as small boys after watching their mum, a self-taught artist. They both specialize in portraits of elephants, rhinos, tigers and other at-risk animals. Many of their paintings are donated to charities to raise funding for animal conservation programmes. Others are owned by private collectors who support their mission – including a member of the UAE's royal family!

Rahul and Rohan have earned various awards over the years, including the Diana Award, the highest honour awarded for youth activism.

HOW TO BE MORE CREATIVE

Creativity is something we all have to some degree, but it's also a skill that can be developed. In fact, some people say that creativity is 10 per cent talent and 90 per cent hard work. So roll up your sleeves and get to it! Here are some tips to give your creativity a boost:

1. **Be curious about everything.** Creative types love to discover new things and are always on the lookout for possible solutions or inspiration. Keep an open mind!

2. **Accidents can be a good thing.** Inventions like penicillin, the Slinky and chewing gum all came about purely by accident.

3. **Play and pretend.** When you play, your mind wanders and your imagination runs free.

4. **Visualize success.** Imagine yourself reaching your loftiest goal and the steps you took to get there. Then go out and try to do that for real.

5. **Leave something undone.** Leaving a task slightly unfinished at the end of a day often makes it easier to get started the next day.

6. **Try something new.** It can be anything. Hula-hooping, juggling, carving wood, archery, whatever. It gets your creative juices flowing.

7. **Hang with people who aren't like you.** Always being around kids who look like you, think like you and act like you is never going to help you E-X-P-A-N-D. Try to meet people with other ways of doing and looking at things.

8. **Draw a picture.** Do it even if you think you can't draw. No one ever has to see it, unless you want them to. Or use chalk to make pavement art!

9. **Make a vision board.** Grab a stack of magazines and rip out any pages that attract your eye. Glue some images to a piece of poster board and put it where you can see it often.

10. **Do a hands-on activity.** LEGO, modelling clay, kinetic sand and slime are all fun materials to mess about with and get you into a creative mood. A great stress reliever too!

11. **Write and record a song** and post it on YouTube. You can always turn the comments off if you're not quite ready for feedback! (Remember you must be at least thirteen and have a parent's permission to sign up to YouTube.)

12. **Take photos** and make an album with them.

13. **Daydream** and let your mind go wherever it wishes.

14. **Unplug.** Take a time out from social media, watching TV, doing stuff on your laptop and wearing headphones or ear buds. As writer Anne Lamott put it, "Almost anything will work again if you unplug it for a few minutes, including you."

15. **Doodle.** Doodling is one of the easiest and simplest activities you can do to boost your creativity – just try not to do it in class!

16. **Paint and colour.** Creating art can reduce stress and improve thinking skills and concentration. Even those who don't consider themselves creative may discover they're an artist at heart!

17. **Write.** Simply writing down your thoughts can really boost creativity. It really doesn't matter what you write. If you can't seem to get started, set a timer for fifteen minutes and just write down whatever comes to mind. Or use this prompt: today, I feel…

18. **Make a video.** It's easy to make a video with a smartphone or tablet. Ask your parents if you can borrow theirs if you don't have one of your own yet.

19. **Tidy your room.** Science has shown that too much clutter kills creativity, although some people say clutter increases it. But there's a difference between 'creative chaos' – surrounding yourself with inspirational items – and simply being a slob. Bonus: this tip will make your parents happy too!

THAT'S NOT FAIR!

It never feels nice when other kids don't want to talk to you or play with you. Or assume things about you, even though they don't know you at all.

Sad to say, this sort of stuff happens a lot, especially to certain people. Some say we should simply accept such things. They shrug and tell us, **"Life isn't fair."**

You know what's wrong with that? Pretty much everything.

It is true that many things in life aren't fair or equal. Those are two different things, mind you. *Equal* is everyone getting the same thing. *Fair* is everyone getting what they need in order to succeed. Sometimes things are not equal and that's fair. Other times things are not equal and that's **NOT** fair. Or right.

If something isn't right, we shouldn't just sit on our hands and do nothing. No matter our age, we all have the ability to stand up, **speak out and do the right thing.**

Equality activists want certain rights to be the same for everyone. They fight against discrimination – that's when someone is treated differently or unfairly because of their race, age, sex or if they have a disability. They also fight against prejudice – having a negative belief or opinion about someone or something that's based on lazy stereotypes and opinions instead of facts. It's very damaging to judge stuff you know nothing about.

No one should be treated badly because of their race, sex, gender, disability, age, religion or who they love. We may look different, believe different things, worship in different ways, eat different food, speak different languages. We may be rich or poor, young or old, this or that or the next thing. But there is one thing that we all need to remember:

"THERE IS ONLY ONE RACE. THE HUMAN RACE."

JANE ELLIOTT

Let's embrace our differences, treat each other with respect and give everyone a fair shot in life. That's what the kids on the following pages are doing and you can do it too!

KIDS TAKING ACTION

DESMOND NAPOLES

Born: 2007 **From:** USA

Known for: speaking out for LGBTQ+ rights

"BE YOURSELF, ALWAYS."

Desmond is a drag kid, actor, model and activist who goes by the stage name 'Desmond is Amazing'. 'Drag' means wearing clothing of the opposite gender, and male drag artists often wear flamboyant wigs and make-up. Desmond made his debut aged seven in a music video and his life has been a whirlwind ever since, with appearances in ad campaigns and *Vogue* magazine, television and strutting the catwalk at New York Fashion Week!

Life can be hard for kids who identify as lesbian, gay, bi-sexual, transgender or queer/questioning and, as an advocate for LGBTQ+ youth, Desmond wants to inspire them – and everyone! – to ignore the haters and be the best they can be.

BENJAMIN BENNETT

Born: 2004 **From:** England

Known for: speaking out for Traveller and Gypsy youth

"GYPSY AND TRAVELLER KIDS ... HAVE THE SAME DREAMS AND ASPIRATIONS AS ANYONE ELSE."

Growing up as a Romany Gypsy has been a challenge for Ben. People have judged or bullied him because of it. Gypsy, Roma and Traveller people have to deal with bias all the time. Ben works hard to show another side and inspire others to stand up to racism.

His role on the TV series *Gypsy Kids* changed many minds about Travellers. His media interviews have also been a force for good. On one TV talk show, he said he dreamed of becoming prime minister. Asked if he'd ever want to live at 10 Downing Street, he said of course – but he'd bring his trailer with him! His fight for equality has earned Ben the Diana Award, the highest honour for youth activism or humanitarian work.

"IT'S IMPORTANT FOR PEOPLE TO STAND UP FOR WHAT THEY BELIEVE IN, AND THAT'S WHAT I'M DOING."

ZULAIKHA PATEL

Born: 2002 **From:** South Africa

Known for: challenging her school's rules

Zulaikha has been stared at, teased and even called a terrorist. Why? Because she is black and sees nothing wrong with wearing her natural hair in an Afro. The style was frowned on at her all-girls high school with teachers saying her hair was too untidy or 'exotic'. Zulaikha felt that having to change her hair was like having to erase her blackness.

Zulaikha was thirteen when she began a peaceful protest at school. Security was called in and she was nearly arrested. The story made headlines and sparked anger across the country. In New York, an artist painted a giant mural of her with a proud halo of hair, pumping her fist. Her school rules were eventually changed and now students can wear their hair how they wish.

SOPHIE CRUZ

Born: 2010 **From:** USA

Known for: advocating for immigrant families

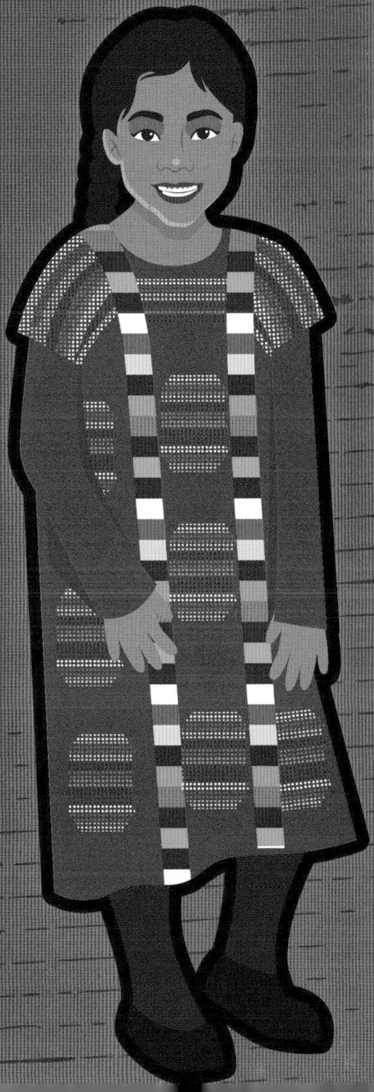

When Sophie was five, the Pope visited the USA. She pushed through the crowd to hand him a letter she'd written in crayon. It said, in Spanish, "My friends and I love each other no matter our skin colour." She hugged him and asked him to talk to political leaders about immigration rights. He did, the very next day.

This story made the news because Sophie was the child of undocumented workers. Her parents had come to America for a better life but didn't have the right to stay. Sophie feared her parents would be sent back to Mexico without her.

Her bold words earned her an invitation to the White House and she went on to win many honours and awards. There's even been a short film made about her life as an activist. Not bad for someone as young as Sophie!

"LET US FIGHT WITH LOVE, FAITH, AND COURAGE, SO THAT OUR FAMILIES WILL NOT BE DESTROYED."

HUNGER HEROES

How many times have you said, "Mum, I'm *starving!*"? (No worries, you're not the only one.) But having the munchies now and then is a far cry from true hunger. The kind of hunger that comes from not having enough to eat for a long, long, long time. Sadly, this is the plight of many people around the world. Even in so-called 'rich' countries.

Without enough food, people don't have the energy to be active or work or pay attention in class. Extreme hunger can also do lasting damage to the body – like blindness, caused by a lack of vitamin A. Millions and millions of people die of hunger every year too.

Across the globe, right at this very moment, about one person out of every nine is hungry. That adds up to an absolutely staggering number of people. And that number is getting larger for all sorts of reasons, including poverty, war and natural disasters.

Another major cause of hunger is climate change. As the earth heats up, the weather gets weirder and weirder and causes storms and droughts and floods. All of these things have an impact on food production.

"IF YOU CAN'T FEED A HUNDRED PEOPLE, THEN FEED JUST ONE."

MOTHER TERESA

Here's the craziest thing. Some people on earth have too little to eat – including people in Great Britain – whilst others have so much that they actually become overweight. Equally mind-boggling: the planet produces more than enough food to feed everyone, yet a third of it is thrown away each year.

What can young activists do about hunger? Some organize class projects such as collections for food banks. Others take food directly to those who need it most. Still others call for completely changing how humans eat, so there's plenty for everyone.

Turn the page to learn about four kids who are taking action in their own amazing way!

KIDS TAKING ACTION

WILL MOORHOUSE

Born: 2007 **From:** Australia

Known for: starting a breakfast club at his school

"[THE CLUB] REALLY HELPS THEM GET TO KNOW EACH OTHER AND THEY'RE ALL SMILES IN THE MORNING."

Breakfast is Will's favourite meal of the day, so he was shocked that kids at his school often came in on an empty stomach. Will knew it would be hard for them to focus, so he had an idea – a breakfast club that offered a free meal every Thursday to any student who wanted one. A local charity donates food and other students prepare and serve it.

The club also makes sandwiches for any kids who might need an additional meal. These are delivered 'under the radar' so no one ever feels embarrassed or singled out. The Breakfast Club has become a weekly social occasion and the students who volunteer get as much out of the experience as the ones enjoying the meals.

CHIARA SACCHI

Born: 2002 **From:** Argentina

Known for: celebrating local food and farmers

"I'M AFRAID THAT MY GENERATION ... WON'T BE ABLE TO SEE THE BEAUTY OF THE WORLD AS PREVIOUS GENERATIONS DID."

Chiara is part of the Slow Food movement, which showcases local farmers, regional cuisine and dining with others. It's all about seasonal, toxin-free food that is good for health, the people who grow it and the planet. But it's a challenge, especially since Chiara lives in an area experiencing the extremes of sizzling summers and bitter winters.

The temperature puts a strain on the electricity grid and regular power cuts make it hard for Chiara to do her homework. The storms are more powerful lately too. One storm ripped the roof off of her house! Chiara joined fifteen other young activists to complain to the UN in 2019, claiming that inaction on climate change is violating children's rights.

SARAH AND CLAIRE JORDAN

Born: 2003 and 2006 **From:** Canada

Known for: co-founding Sarah & Claire's Food Drive

"CITIES CAN DEFINITELY MAKE A DIFFERENCE. COMMUNITIES CAN TOO . . . IT'S ABOUT AWARENESS."

One night during dinner, six-year-old Sarah said she was full, even though there was still food on her plate. Her mother told her how fortunate they were and, moments later, a story came on the radio about a Thanksgiving food drive. Sarah announced that she wanted to do something to help and started a food drive of her own. Her sister Claire got involved too, so they called their project Sarah & Claire's Food Drive.

Over the past eleven years, the sisters have helped provide over 500,000 meals to people in need and have inspired over 100,000 young people in Toronto to get involved with their community. Sarah and Claire were both given the Diana Award in 2019.

GENESIS BUTLER

Born: 2007 **From:** USA

Known for: founding Genesis for Animals

Genesis was three when she found out where chicken nuggets came from. She stopped eating meat immediately and later gave up milk and cheese as well. By ten, she was one of the youngest ever TEDx speakers, delivering a talk on healing the planet with a plant-based diet. Genesis knows her stuff! She also started a foundation, Genesis for Animals, which gives money to animal sanctuaries and rescue groups.

In 2019, Genesis teamed up with the Million Dollar Vegan campaign and challenged Pope Francis to go vegan for Lent! If he did, $1 million would be donated to the charity of his choice. The Pope didn't go for it but he did give her a special blessing. Genesis still considers the campaign a success, as it helped inspire many others to try living a cruelty-free life.

"BY CHANGING WHAT YOU CHOOSE TO EAT, YOU WILL BECOME A PART OF HEALING OUR PLANET."

THE BLUE EYES BROWN EYES EXPERIMENT

For centuries, people have faced racial prejudice, which is judging someone because of their culture or the colour of their skin.

During the 1950s in the USA, the civil rights movement began to try to end racial discrimination and ensure that black Americans had the same rights as white people. In spring 1968, one of the leading campaigners, Martin Luther King Jr, was shot dead. The next day, a teacher in a small American town decided to try an experiment with her young students, all of whom were white. She wanted them to understand how racial discrimination worked.

"HOW DO YOU THINK IT WOULD FEEL TO BE A BLACK BOY OR GIRL?" SHE ASKED. "IT'S HARD TO KNOW, IF YOU'VE NEVER EXPERIENCED PREJUDICE OR DISCRIMINATION YOURSELF."

The teacher separated the blue-eyed kids from the kids with brown eyes. She had the blue-eyed children put on paper armbands. Then she told everyone that people with brown eyes were smarter. She let the brown-eyed children eat lunch first and gave them a longer recess. Only the brown-eyed children were allowed to drink from the water fountain.

The mood in the classroom totally changed. The brown-eyed kids acted more confident. They were also meaner and said nasty things to the blue-eyed kids. The students with blue eyes started making mistakes and seemed timid and sad. The two groups stopped playing together and fights broke out.

The next day, things were reversed. This time it was the blue-eyed students who were smarter. They ate lunch first and had a longer recess and drank from the water fountain. But something was different. The blue-eyed kids weren't as mean as the brown-eyed kids had been. Why not?

The blue-eyed kids explained that it felt horrible to be on the bottom. They didn't want to make anyone else feel like that. **Ever.**

This experiment is still being talked about more than fifty years later. It's changed how many people think about prejudice and discrimination. It's inspired them to speak up when they hear or see something that's wrong. Hopefully **YOU** will do the same!

HOW TO FIGHT UNFAIRNESS

The best way to fight something that's wrong is to do something about it. Some issues, like racism or prejudice, are not easy things to fix. But there are many other situations in life that are out of whack in some way – one of those might be a good place for a young activist to start! Here are some questions to think about:

IS IT REALLY UNFAIR?

Some things, like rain on your birthday, are disappointing but not unfair. They're beyond our control. Other things, like losing a football match, are frustrating but not unfair either. But what if boys always get called on in class and girls don't? What if a park in a rich neighbourhood is always kept nice and tidy, but a park in a poor neighbourhood is ignored? These things are NOT fair and something should be done about them.

WHAT WOULD YOU DO?

Complaining without suggesting a solution won't accomplish much. If something is unfair, figure out what would be fair.

WHO'S IN CHARGE?

Next, figure out who has the power to make decisions and change the situation. Maybe it's a teacher, a coach or a local MP. Or maybe it's your parents.

HOW DID IT GET TO BE UNFAIR?

Politely ask why things are the way they are. Just because something's unfair doesn't mean it was done on purpose. Maybe no one ever noticed it was unfair until you spoke up. Or maybe they will explain their decision and you'll come to see things differently. Whatever happens, keep your cool. People can get defensive when they're questioned (especially by a kid).

WHO ELSE CAN GET INVOLVED?

If there are others concerned about the same issue, it's time to join together. Speak out with petitions, letters or peaceful protests. The more attention you get, the more pressure there will be on those in charge.

ARE YOU WILLING TO GO THE DISTANCE?

Still not getting anywhere? It may be time to take matters into your own hands. Let's say tickets to a school dance are too pricey for some kids. Plan an event everyone can afford. If someone at your school is being picked on, invite that person to sit with you and stand up to the bullies.

DON'T JUST SIT THERE. YOU'RE AN ACTIVIST. TAKE ACTION!

ECO -PRENEURS

Breaking news: you don't need to be a grown-up to start your own company. An entrepreneur is someone who starts their own business. An eco-preneur is someone who starts an earth-friendly business. All it takes is brainpower, a load of hard work and a desire to do something good for the planet.

Could YOU be an eco-preneur? Take this just-for-fun quiz to see if you might be a mogul in the making:

Q: ARE YOU A RISK-TAKER?

Starting a business is never a sure thing but entrepreneurs can deal with that. Taking risks is part of being a fearless leader.

Q: ARE YOU CURIOUS?

Entrepreneurs tend to have a strong sense of curiosity. They LOVE asking questions. What if there was a better way to…? Why doesn't anyone ever try…? What would happen if…? They are always thinking up new ways to solve problems.

BIZ WHIZ KIDS

According to one recent poll of young people aged ten to seventeen:

- 77 per cent want to be their own boss
- 45 per cent hope to start their own biz
- 42 per cent plan on inventing something to change the world

Q: CAN YOU HANDLE BUMPS IN THE ROAD?

Entrepreneurs don't call it quits when the going gets tough. They are determined. They want their project to succeed!

Q: CAN YOU BE A CHAMELEON?

Chameleons can change colours, which is pretty awesome. While you don't need to turn green to run a business, you DO need to be able to change if necessary.

Q: DO YOU HAVE A COMPETITIVE NATURE?

Everyone wants to succeed but eco-preneurs are like a dog with a bone. They. Never. Give. Up!

If your answers were mostly **YES**, there could be an eco-business in your future. Read on about four kids who have already made the leap…

KIDS TAKING ACTION

OSKAR PETERSEN

"EVERY PIECE OF PLASTIC EVER CREATED IS STILL ON THE EARTH TODAY!"

Born: 2009 **From:** England
Known for: founding OSKeco

When Oskar saw a photo of a sea turtle with a plastic straw in its nose, it was a life-changing moment. Everywhere he looked he saw plastic and no wonder – around the world, one million plastic bottles and two million plastic bags are produced every minute. He wanted to give people better options. He did some research and asked his parents to invest in his first business idea – making and selling bamboo toothbrushes.

Oskar has since expanded his online shop, OSKeco. He now sells reusable drinking straws, water bottles, bamboo cutlery sets and cotton shopping bags, too. And his company logo has a sea turtle on it!

MIKAILA ULMER

"DON'T GO INTO BUSINESS TO MAKE A LOT OF MONEY. CREATE A BUSINESS THAT YOU SEE THE THAT WORLD IS MISSING."

Born: 2005 **From:** USA
Known for: founding Me & the Bees Lemonade

It started when bees stung Mikaila twice in one week. Soon after, her great-grandmother sent her a cookbook with old-fashioned recipes that used honey. It gave Mikaila an idea – a stand selling lemonade sweetened with local honey. She quickly went on to sell her products at Whole Foods and created more buzz when she went on *Shark Tank* (the US *Dragon's Den*). She was only ten but came away with $60,000 to expand her business.

Mikaila sells 360,000 bottles of her lemonade each year and donates a percentage of her profits to organizations that protect bees, which are vital to our ecosystem. She is one of America's youngest business owners and has the best job title ever: Queen Bee!

MAX LAWTON

Born: 2001 **From:** New Zealand

Known for: co-founding ClearFreeSkin

"INNOVATION IS THE KEY TO SUCCESS IN THE MODERN WORLD."

Max suffered from really bad acne for most of his teenage years. He tried everything to treat it but didn't like using medication or products made with lots of chemicals. Since he couldn't find an all-natural product on the market, he decided to create his own.

Max and two friends started a skincare company, ClearFreeSkin, after researching manuka oil. New Zealand's native Maori people have used this 'miracle' ingredient for centuries. It fights bacteria and soothes red and irritated skin – the perfect thing for acne. The company's first product – an easy-to-apply facial spray in a glass bottle – was a hit. Will now runs the company on his own and sells products worldwide. A clear success!

LUCA BERARDI

Born: 2005 **From:** Italy

Known for: founding Young Animal Rescue Heroes (YARH)

Multi-talented Luca is a lot of things: singer, actor, author, TV host, entrepreneur and conservationist. After reading about endangered species he wanted to do something to help.

When he was eight, Luca founded his own environmental organization to make money for animal charities. While living in Kenya, he collected wastepaper and old juice cartons from schools and offices and sold them to recycling companies.

Now based in Sweden, Luca continues to donate the money YARH earns to organizations that fight poaching and protect wildlife habitats. He's also given a TED Talk, hosts environmental workshops and organizes community clean-ups. What will Luca do next? Watch this space!

"I WISH THAT EVERYONE WANTED TO DO SOMETHING FOR THE ENVIRONMENT AND FOR ENDANGERED WILDLIFE."

START YOUR OWN ECO-COMPANY!

There's never been a better time to start a 'green' business. Here are some tips for how to do well by doing good.

BRAINSTORM

Look at your hobbies and interests as a starting point. Decide what you want your company to be about and which product or service you can sell or provide.

FIND A USP

A USP is your 'unique selling point'. What will set your company apart from all the others out there?

SEEK INSPIRATION

Think about some of your favourite earth-friendly brands and what you like most about them.

CREATE A PLAN

How will you get supplies? How will you share the news about your product? Talk to your parents or another adult about things you'll need their help with, like shipping, marketing and setting up a website.

SELLING ONLINE

If you want to set up an online shop on Etsy, Folksy or a similar site, ask your parents for permission first. They will be able to help you with managing your shop, marketing, setting prices and shipping items!

GET FUNDING

Having an idea is one thing, but you'll need money to get up and running. You might consider crowdfunding, which raises cash from friends, family, customers and potential investors.

NETWORK

Speak to other green biz owners and ask questions. Give yourself a crash course in business ownership and learn as much as you can. Chances are someone knows someone who can help you with … whatever. You all have the same goal, after all: to do something good for the planet!

PLAN A LAUNCH AND INVITE EVERYONE YOU KNOW

CONGRATULATIONS!
YOU'RE IN BUSINESS!

FUN IDEAS FOR AN ECO-BIZ

The sky's the limit when it comes to creative ways to make money. Here are some suggestions to get your eco-biz brainstorming session started.

DRINKS STALL

In the United States, young people often sell homemade drinks at a 'lemonade stand'. Is there a market for something similar in your neighbourhood?

CAKE DECORATING

Offer your services to friends and family who are hosting special events.

JEWELLERY-MAKING

There are tons of creative ways to up-cycle something old into something new, like beach glass. Look on Etsy for ideas or other online DIY sites.

FASHION DESIGN

Have an eye for fashion? You can design your own line or up-cycle charity shop finds.

ORIGINAL ART

Sell your paintings and drawings at local arts and crafts fairs or set up an online shop.

FACE-PAINTING

This is always a hit with younger kids and it's fun to do too! Offer your services at school fairs or other special events like birthday parties.

SECOND-HAND BOOKS

Collect books from friends and family and sell them on Amazon or other selling sites.

GIFT-WRAPPING SERVICE

Offer 'green' gift-wrapping services during the holidays. Use recycled paper or newspaper and add natural accents like small pinecones or holly leaves.

CAR BOOT AND TABLE-TOP SALES

There are lots of people who'd love to sell some of their stuff but are too busy to do it. You can offer to sort and display items, set prices and collect money, either for a flat fee or part of the profits.

PARTY PLANNING

You can help plan earth-friendly parties and special events, with things like vegan cake and green alternatives to balloons and plastic.

GREETING CARDS

If you are artistic, you could design and sell cards. It's fun to do and a great way to recycle and up-cycle paper!

LAWN CARE

If you enjoy being outside, start a mowing service, or help people tend to their gardens by raking leaves or shovelling snow. Doing things the old-fashioned way, without leaf blowers or snowblowers, is much better for the planet!

RECYCLING

Set up a recycling service where you collect things like paper, bottles and cans from people who can't be bothered to do it themselves.

COSTUME-MAKING

Do you have a knack for making your own fancy dress and Halloween costumes? Make a business out of it by designing costumes for other people as well.

ORGANIC SOAP

Making soap isn't as hard as some people think and there are lots of ways to make it unique. You can make goat's milk soaps, organic soaps, herbal soaps…

CANDLES

Regular candles often release harsh or even toxic chemicals into the air, but soy and beeswax candles do not. They also burn slowly and without any smoky residue. Candles make great gifts and you can also sell them at craft shows or online.

REFURBISH BICYCLES

One way to reduce carbon emissions is to get more people to ride bikes. If you like fixing things, this would be a great service to offer.

HOMEMADE BATH AND BODY PRODUCTS

More and more people are interested in all-natural products. You can make scrubs using ingredients such salt, coconut oil and essential oils.

HAPPY. HEALTHY. HERE.

Mental health is a hot topic, but what exactly is it? **When people talk about mental health, they mean your emotional well-being** and how you cope with everything life throws at you, which is sometimes called 'resilience'. Just like your physical health, sometimes your mental health is good and sometimes it's not so good. Think of it as your emotional fitness!

When our mental health isn't great, we may experience things like depression, anxiety, eating disorders and panic attacks. Young people struggle with these issues just as much as adults. **The best way to deal with such feelings is to face them head-on.**

Mental health activists want everyone to know that we all have troubles – even those whose lives seem perfect on the outside. (Maybe especially those people.) This kind of activism is all about sharing info, raising awareness and giving others the support they deserve.

There is one mental health issue that's becoming more and more common among young environmental activists: **eco-anxiety.**

Greta Thunberg developed eco-anxiety after she learned about global warming in school. Unlike her classmates, she couldn't stop thinking about it. The future of the planet worried

YOU CAN'T POUR FROM AN EMPTY CUP – TAKE CARE OF YOURSELF FIRST.

her so much that she became very depressed. So depressed that she stopped eating and speaking for a while.

Greta finally told her parents about her fears and anxiety. It made her feel a lot better to talk about it. To her surprise, her words convinced her parents to make some big changes, too. They gave up meat, put solar panels on the house and quit travelling on aeroplanes. Greta's mum even gave up performing as an opera singer, as it required flying internationally. It was only then that Greta realized she could make a difference by inspiring others to take action.

Having depression is extremely hard, but focusing on her climate activism became Greta's path to recovery. It's what motivated her to make a hand-lettered sign that read 'School Strike for Climate'. The rest is history!

KIDS TAKING ACTION

HAILEY HARDCASTLE

Born: 2001 **From:** USA

Known for: pushing for mental health days

"IT'S HARD WHEN ADULTS INVALIDATE THE STRUGGLE THAT TEENS GO THROUGH."

Hailey was under a lot of stress when she started having a panic attack one day in class. Panic attacks can be super scary and her mum got her out of school by saying she had a doctor's appointment. Having some time to chill out at home did Hailey a world of good. It made her wonder why anyone should have to lie if they needed a 'mental health day'.

Students at Hailey's school were only allowed to miss class because of illness, appointments or emergencies. But Hailey knew kids took 'sick days' when they were having a hard time dealing with their emotions. This only made things worse – since adults were unaware of the real issue, kids didn't get the help they needed. Hailey took action by helping draft a mental health day bill, which was signed into law!

LYLA-ROSE O'DONOVAN

Born: 2013 **From:** England

Known for: sending certificates of bravery

"HELP MAKE SOME SMILES!"

For most of her life, Lyla has suffered a great deal of pain. She was diagnosed with a brain tumour when she was only three years old. Without an operation to remove it, doctors said she might have only days to live. She made it through that surgery – and many others – and her determination to live and love has been an inspiration to all.

Lyla set out on a mission to bring joy to other kids battling illness. With the help of her parents, she set up a Facebook page called 'Lyla and Lilley's Stars'. She and her older sister, Lilley, send out certificates of bravery to children all around the world. In 2019, she was honoured with a WellChild Award and met The Duke and Duchess of Sussex.

RAINA IVANOVA

Born: 2004 **From:** Germany

Known for: speaking up for kids' rights

"[THE ADULTS] DIDN'T REALLY HAVE TO FACE ALL THESE PROBLEMS AND ISSUES THAT WE HAVE TO FACE NOW."

Germany was so hot in the summer of 2019 that it set a new record: 40.5°C (105°F). It made Raina's classroom unbearable and dried up Hamburg's famous canals. Her city has also experienced the opposite extreme – heavy rain and intense storms that caused flooding.

The fact that such wacky weather is caused by climate change has taken a toll on Raina's mental health. Worrying about global warming has disrupted her life. She gets emotional whenever she thinks about what's happening to the planet, as it's kids like her who will be the most impacted. Raina was one of sixteen children who presented an official complaint to the UN to protest lack of government action on the climate crisis.

AMANDA SOUTHWORTH

Born: 2002 **From:** USA

Known for: coding for coping skills

Amanda was bullied at school for many, many years. What hurt the most was that people saw what was happening yet did nothing. Clueless about where to go for help, she secretly downloaded a psychology textbook and taught herself how to cope with her own anxiety, depression and PTSD (post-traumatic stress disorder). That inspired her to create an app for other kids dealing with the same problems.

AnxietyHelper is a sort of guidebook for mental health and now has thousands of active users and followers on social media. Amanda was selected as one of *Teen Vogue*'s '21 Under 21'. She was so motivated to help other people that she opened Astra Labs, her own app development company.

"IF YOU SEE SOMETHING WRONG, FIX IT. IF YOU SEE SOMEONE HURT, HELP THEM."

ARE YOU ECO-ANXIOUS?

Eco-anxiety is the real deal. It's when you are stressing out over climate change and feel worried about what it's doing to the planet. If you feel this way, it's because you're smart and compassionate and you care. You're probably better informed than many adults, too.

Eco-anxiety is more than just being sad about the plight of polar bears. It's about losing sleep over melting ice caps and greenhouse gases and crazy weather. It's having so much fear or feeling so hopeless that it starts messing with your daily life. It's feeling guilty, angry, sad, numb, disappointed, depressed, stressed, lonely, tired. Maybe all of these things. Not good.

Some adults say that kids should be kids and not worry about climate change. **WRONG.** Generation Hope is already well aware of what's going on in the world – it's impossible to hide reality from anyone old enough to take a good look around or surf the internet.

Just because you're young doesn't mean your emotions are not 100 per cent legit. Your feelings shouldn't be brushed aside with responses like, "Don't be silly!" or, "You're just a child!"

The good news is that you can turn eco-anxiety into a force for change. Because in Greta Thunberg's words, **"It is still not too late to act."**

AVOIDING ACTIVIST BURNOUT

Being an activist can be hard. Ridiculously hard. But you can't make a difference if you are too overwhelmed or worn out to care any more. As flight attendants say on aeroplanes, "Put your own oxygen mask on before helping others." Excellent advice!

Here are some tips to avoid activist burnout:

- **Set the bar lower.** Sometimes it's okay to be just 'good enough'. It's impossible to be a high achiever 24/7.
- **Stay healthy.** Get enough sleep and exercise, scrap the junk food and drink lots of water – preferably from a reusable bottle!
- **Do something you love every day.**
- **Have a sense of humour.** This quality will take you very far in life.
- **Just be.** Find a place where you can just be however you are. No fake smiles or pretending to be someone you're not. A yoga class, a dark cinema … whatever works for you.
- **Ask for help.** Successful activists know when they need assistance and advice.
- **Have other interests.** Not everything in your life has to be devoted to activism!
- **Do less.** Doing something every single minute of the day is a recipe for burnout. We are human be-ings, not human do-ings. Cut back and don't feel guilty about it.
- **Don't live for likes.** Doing things just to please others is a total waste of time. Celebrate your own achievements, however small.
- **Keep a journal.** If you ever start feeling overwhelmed, write down what's on your mind. Just putting it all down on paper can help.
- **Get out in nature.** Spending time outdoors can be so healing and energizing.
- **Team up.** By joining forces with others, you can do more and have a greater impact.
- **Talk it out.** If you feel like venting, vent. Holding stuff in can take a toll on your health. If you don't feel like talking to a friend, try talking to your pet instead. Pets are the world's best listeners. They give you their full attention, aren't judgemental and never interrupt!

WATER WARRIORS

Water, water, everywhere … we use it for all sorts of things, from cooking and drinking to growing food to washing. We all need water to survive and it's something many of us take for granted.

But maybe we shouldn't. A lack of clean water is a big problem in many places around the world. Can you imagine:

- SPENDING HOURS EACH DAY FETCHING HEAVY JUGS OF WATER?
- TURNING ON THE TAP AND HAVING DARK, BAD-SMELLING WATER TRICKLE OUT?
- GETTING A SKIN RASH AFTER TAKING A SHOWER IN DIRTY WATER?
- SEEING MOUNDS OF TRASH FLOATING DOWN A BEAUTIFUL RIVER?
- WATCHING CROPS DIE DUE TO DROUGHT?
- HAVING TO FLEE FROM RISING FLOODWATERS?

Believe it or not, such things happen to people on a daily basis, including young people like you.

HEY, DON'T FLUSH THAT!

- medicine
- wet wipes
- cotton balls
- pet poo
- food waste
- towels, liners and tampons
- nappies

This stuff could get into the water system and harm animals!

DID YOU KNOW?

- A person can live about a month without food, but only three to four days without water.
- Although 70 per cent of earth is covered by water, most of it is salty or frozen.
- If all the world's water could be poured into a gallon jug (3.8 litres), only one tablespoon would be fit for human use.
- One in three people globally do not have access to safe drinking water.
- In some countries, people walk an average of 6.5 km (4 miles) a day to get water.
- A leaking tap can waste up to 3 litres (3 quarts) of water a day.
- Bottled water may not be as safe to drink as regular old tap water.
- Freshwater animals, such as river dolphins, are vanishing faster than those on land or at sea.
- A full bath uses about 80 litres (21 gallons) of water. A five-minute shower uses 40 litres (10.5 gallons).
- Every year, 300 to 400 million tonnes of poisonous waste are dumped into our waterways, made up of toxic sludge and dangerous chemicals, amongst other things. **EWWW!**

Turn the page to find out what four activist kids are doing to help!

KIDS TAKING ACTION

"WE WILL CONTINUE TO TAKE TO THE STREETS UNTIL CONCRETE MEASURES AGAINST THIS CRISIS ARE IMPLEMENTED."

ARAN COSENTINO

Born: 2003 **From:** Italy

Known for: fighting to save an Italian river

Aran became an activist aged twelve, when he started a school recycling project. A couple of years later, he heard about plans for a new hydroelectric power station on the Alberone River. Aran knew this project would take a terrible toll on one of Italy's last pristine rivers, which is home to freshwater crayfish and other wildlife. The power station would put these creatures in danger and harm the area's natural beauty as well.

Aran joined other activists to stop the project, getting the word out on social media about why it was so important to defend the river. The battle took two long years but the company finally dropped the project. The activists won! Aran now participates in weekly school climate strikes and hopes to meet Greta Thunberg someday.

AUTUMN PELTIER

"WATER IS LIFE. MOTHER EARTH DOESN'T NEED US, WE NEED HER."

Born: 2004 **From:** Canada

Known for: being a 'water warrior' for the Anishinabek nation

Autumn is a Native American who became an activist when she was eight. She has travelled the world with her message that water is sacred and must be protected and has taken part in important UN events.

One of her proudest moments came when she presented a ceremonial water bowl to Canada's prime minister, Justin Trudeau. He had supported oil projects in wilderness areas and the bowl was to remind him of his duty to protect nature. Autumn handed over the bowl and said she was very unhappy with the choices the prime minister had made. That was probably the last thing the prime minister expected to hear from a twelve-year-old girl!

"THIS IS OUR WORLD, NOT, 'I HAVE MY COUNTRY. YOU HAVE YOUR COUNTRY'."

AYAKHA MELITHAFA

Born: 2002 **From:** South Africa

Known for: being a spokesperson for the African Climate Alliance

Ayakha has seen the impact of climate change with her own eyes, like the droughts that have taken such a terrible toll on southern Africa. Her mother can't plant crops due to lack of rain and some of their farm animals have died. Ayakha became so frustrated by the state of the world that she joined fifteen other kids – including Greta Thunberg – to file a complaint to the UN. The charge? Failing to take action on climate change, which they say is violating the rights of children.

Ayakha wants people to know that global warming affects everyone, not only those who can afford an eco-friendly lifestyle. She believes it's really important for the poor and people of colour to speak out. The time has come for all of us to stand up for the planet!

AMARIYANNA 'MARI' COPENY

Born: 2007 **From:** USA

Known for: demanding clean water for Flint, Michigan

When Mari was eight, her hometown had been without clean water for over four years. The drinking water had high levels of bacteria and lead. It smelled bad and looked brown and disgusting. No one wanted to drink it or bathe in it and those who did were often unwell – some even died.

Well, enough was enough and Mari had to do something. She wrote a letter to US President Barack Obama and told him about her town's water problems and asked to meet him. Her heartfelt words inspired him to come see what was going on for himself. The president gave the city $100 million to repair its water system. Mari still works as an activist and hopes to run for US President in 2044. Clean water will be one of her top priorities.

"MAKING ADULTS REALIZE I AM SERIOUS IS THE BIGGEST CHALLENGE."

18 WAYS TO SAVE WATER

1. **Don't flush.** "If it's yellow let it mellow, if it's brown flush it down." This rule doesn't apply when you are out in public or at other people's houses!

2. **Re-wear those clothes.** Check your clothes each night and re-wear anything that is still mostly clean. Only toss things in the wash when they are good and dirty!

3. **Bathe less.** You only need a bath once or twice a week, unless you're really dirty.

4. **Reuse bath water for plants.** Don't let your bath water go down the drain. Scoop it up with a bucket and use it to water houseplants, your garden and trees.

5. **Take showers.** Showers use much less water than baths, especially if you're speedy.

6. **Reuse your towel.** Use the same towel for a whole week. After all, you should be clean after a shower! If you hang your towel to dry it will be all ready to go next time.

7. **Turn off the tap while brushing your teeth.** That will save litres of water every day!

8. **Be a leak detective!** Listen for dripping water and let someone know if you spot a leak.

9. **Fill the dishwasher.** Dishwashers use less water than doing dishes by hand. Fill the dishwasher all the way up before running it. Empty it too to score points with your parents!

10. **Use the same glass.** Don't grab a new glass every time you're thirsty. Use one and keeping refilling it throughout the day. Fewer dishes to wash means less water wasted.

11. **Pour what you need and no more.** It's better to drink half a glass and go for a top-up than pour a whole glass and tip some away!

12. **Keep cold water in the fridge.** Running the tap for cold water is wasteful. Keep a pitcher of cold water in the fridge instead.

13. **Don't toss clean, unused water.** Instead, use it to water plants or refill your pet's bowl.

14. **Wash fruit and veg in a bowl of clean water.** Use a bowl instead of washing produce under running water.

15. **Grow your own food.** Gardening is fun and it's less wasteful than watering a grassy lawn.

16. **Give up the water guns.** Fun, yes. Wasteful, yes.

17. **Collect rainwater.** Ask your mum and dad to set up a water butt to catch rainwater.

18. **Ditch the hose.** Save water by washing your bike with a bucket and cloth, rather than using the hose.

HOPE
NOT
HATE

"HOPE IS NOT SOMETHING YOU HAVE. IT'S SOMETHING YOU CREATE, WITH YOUR ACTIONS."

ALEXANDRIA OCASIO-CORTEZ

After a mass shooting at their Florida high school in 2018, student survivors organized walkouts at dozens of schools across the US to protest against gun violence. Later, more than a million people joined their March for Our Lives in the nation's capital. It was a peaceful, non-violent protest – a plea for the world to ban guns and choose hope over hate.

These student-led protests inspired Greta Thunberg to do something similar in Sweden to protest climate change. Look how that turned out! She's now the teen general of a growing global army of youth activists. **One small action can turn into something HUGE.**

Peace activists are involved in a wide range of issues. They are working to end wars, stop violence, battle racism and speak out against bullying and hate crimes. As they see it, the earth is a small planet with limited resources. **We need to co-exist in peace and harmony. We all need to get along and work together to solve the planet's problems.**

Our world needs peace and hope more than ever, but where should we start? The best advice is to start small. Start local. Changing what's right around you – your home, your school, your community – might lead to changes on a larger scale than you ever imagined.

Leaders like Mahatma Gandhi, Nelson Mandela and Martin Luther King Jr all believed that we can change the world with dignity and love, not hate. In Gandhi's own words, **"The future depends on what you do today."**

KIDS TAKING ACTION

HAVANA CHAPMAN-EDWARDS

Born: 2011 **From:** USA

Known for: protesting against gun violence

"EVEN IF YOU'RE TINY, YOUR VOICE IS NOT."

Havana first gained media attention at age seven, when she was the only child at her school to walk out to protest gun violence. She wanted to stand up for African American victims of gun crime, like her cousin who died while still in his teens. Havana was disappointed that her classmates weren't as outraged about gun crime as she was.

She pulled on a bright orange astronaut suit – the colour is worn in memory of victims and survivors of gun crime – and had her mum take a photo. They posted it on Twitter with the caption, "I am all alone at my school but I am not alone." After her tweet went viral, Havana was invited to speak at a big rally in Washington, DC, and was featured in *TIME* magazine's 'Guns in America' issue.

TOKATA IRON EYES

Born: 2003 **From:** USA

Known for: speaking out for indigenous rights and culture

"WE HAVE TO REALIZE THAT WE'RE ALL ON THE SAME TEAM, AND WE'RE ALL FIGHTING FOR THE SAME THING."

Tokata met Greta Thunberg at a climate crisis meeting at the Pine Ridge Reservation in South Dakota, USA. Tokata felt they were kindred spirits and invited the Swedish teen activist to visit her ancestral homelands.

Welcoming a blue-eyed European to lands that many Native Americans consider sacred was a pretty big deal. These lands have been pillaged and destroyed by white people for centuries and tribal nations have had little to no say about it.

Tokata has been an activist since she was nine, when she first spoke out against uranium mining, and her newfound friendship with Greta sent a clear message – she's one of us.

JOCELYN GOLDSTEIN AND ELSIE LUNA

Born: 2007 and 2009 **From:** England

Known for: co-founding XR Kids

"[XR KIDS IS] STUDENT RUN. THERE ARE NO ADULTS INVOLVED."

When Jocelyn first heard about Greta Thunberg's solo protest in Stockholm, she realized that anyone could fight climate change. She got involved with Extinction Rebellion – XR for short – inspired by the group's peaceful (and often controversial) demonstrations around the UK. She also became friends with Elsie Luna after they met in XR Youth, a sub-group for young eco-activists.

But there were no other activists their age in XR Youth, so, Jocelyn and Elsie figured they might as well start their own group. XR Kids (XRK) is still quite new but already has thousands of followers on social media.

BANA ALABED

Born: 2009 **From:** Syria

Known for: tweeting in the midst of war

Bana's hometown of Aleppo was one of the main battlefields of Syria's civil war. Her home was bombed while she was playing inside. Then her best friend was killed and children weren't allowed to go to school any more.

In late 2016, Bana started posting on Twitter. With her mum's help, her tweets documented the air strikes and the hunger and fear, capturing hearts around the world with her dreams of a normal, peaceful childhood. Bana and her family eventually escaped to Turkey, where she wrote a book, *Dear World: A Syrian Girl's Story of War and Plea for Peace.*

Bana received a Diana Award in 2019 and continues to be an advocate for children's rights and world peace.

"I WANT THERE TO BE NO MORE WARS IN THE WORLD SO EVERY CHILD CAN HAVE THE CHANCE TO DO ANYTHING THEY WANT."

25 WAYS TO MAKE PEACE

1. Sometimes in life, it's best to agree to disagree and move on.

2. Be kind and compassionate whenever you can.

3. Practise yoga or meditate to calm your mind.

4. Write an article about peace for your school newspaper.

5. Post an inspiring quote about peace on social media.

6. Attend a peace rally.

7. Listen to relaxing music or nature sounds.

8. Practise speaking another language or watching a film with subtitles.

9. Refuse to watch violent movies or play violent video games.

10. Make a positive difference by volunteering.

11. Write to a local or national political leader.

12. Start a collection to donate to a charity.

13. Consider someone's intentions rather than the outcome.

14. Step in when you see someone in trouble.

15. Take a break from watching or reading the news.

16. Share stories about positive things around the dinner table or at school.

17. Turn your bedroom into a calm and tidy retreat from the world.

18. Wear a T-shirt with a message about peace or hope.

19. Try your best not to judge others.

20. Read books about peace or peace-makers.

21. Forgive others, even when it's really hard to do so.

22. Create art or poetry or music with peace as the theme.

23. Visit museums or travel to learn about life in other places.

24. Organize a peace-focused community project, like painting a group mural.

25. Say, "I'm sorry" – and mean it!

TALKING ROCKS ROCK!

Next time there's a tense situation at home, use a Talking Rock (or Talking Stick) to keep the peace.

Find a pretty rock and maybe decorate it with paint, markers or ribbons. Whoever holds the rock gets to speak without interruption. Then hand the rock to the next person to speak. Sounds crazy but it really helps!

PEACEFUL PROTESTS FOR BEGINNERS

Boycotting. Refusing to buy or use a certain item or service. For example, boycotting a circus because it features animal acts.

Demonstrations. When a large group of people come together to show they are unsatisfied about something. For example, the March for Our Lives in Washington, DC, to protest for tighter gun control.

Hashtag activism. Using social media to share important activist information. For example, the #FridaysforFuture movement.

Media relations. Using the media – newspapers, magazines, TV, internet and radio – to raise awareness about a cause. For example, activists giving interviews at student strikes.

Petitions. Gathering signatures on paper or online to show decision-makers how many people support a certain cause.

Strikes. Refusing to work or go to school. For example, Greta Thunberg's solo school strike in Sweden, which has since become an international movement.

SHELTER SUPERSTARS

There are all sorts of reasons why people end up without a home. It might be because of a natural disaster, a war or the loss of a job. They might not have enough money to pay the rent or have anyone who can give them a helping hand.

A homeless person might be a former soldier with no place to go after leaving the military, a person who has escaped from a harmful situation or someone who has mental or physical health problems. All too often, homelessness is not a choice but a journey that many people find themselves on through no fault of their own. It's a complicated and very misunderstood issue.

For most of us, it's almost impossible to imagine not having a home or having to struggle every day to survive. Sadly, this is the plight of millions of people all over the world – including kids who might be much younger than you. **The things so many of us take for granted, someone else is praying for.**

The question is … what can be done about poverty and homelessness?

One thing we can all do is to have a sense of empathy and compassion. Every person living on the street or in a shelter has a unique story. They need help, not judgement or blame. All human beings – homeless or otherwise – deserve to be treated with dignity and respect. That's what the young activists on the following pages are doing, each in their own special way. Join them!

YOU (YES, YOU!) HAVE THE POWER TO CREATE A BETTER WORLD FOR ALL

KIDS TAKING ACTION

JAHKIL JACKSON

Born: 2008 **From:** USA
Known for: founding Project I Am

"IN A PERFECT WORLD, I WOULD BUY EVERY HOMELESS PERSON A HOUSE."

Helping his aunt distribute food to the homeless in Chicago was an eye-opening experience for Jahkil. He wanted to do more – a lot more. He started off by making Blessing Bags filled with basic supplies. His project expanded to other US cities as well as to orphans and victims of natural disasters in Swaziland, Guatemala and Puerto Rico. By 2019, he'd given away more than 20,000 bags.

Jahkil became a motivational speaker, urging other kids to raise awareness about homelessness. He serves as a Youth Ambassador for a global anti-poverty organization and joined the WE International Youth Council. Jahkil's biggest wow moment (so far) was being honoured by former US president, Barack Obama.

MALACHI JUSTIN

Born: 2009 **From:** England
Known for: inspiring Project Malachi

"DEAR SALVATION ARMY, HERE IS £5 I GOT FROM THE TOOTH FAIRY. PLEASE USE IT TO BUY HOMES FOR PEOPLE WHO ARE HOMELESS."

Malachi lives in a part of London with a large population of homeless people. He became very concerned about them but didn't think there was much that he could do. Or could he?

One day, he posted a letter and a £5 note to The Salvation Army charity and asked them to help buy homes for homeless people. That humble donation kickstarted something BIG – a £5 million hostel built using forty-two converted shipping containers! Each unit is designed like a studio flat, with a bed, sofa, bathroom and kitchen area.

The site also hosts 'Recycles' – a workshop that trains young people to repair and sell on old bicycles. It's amazing what one small act can accomplish!

SHIREEN AND HASAN ZAFAR

Born: 2003 and 2001 **From:** Pakistan

Known for: starting a school for street children

"WE BELIEVE THAT EVERYONE DESERVES TO BE ALL THAT THEY CAN BE."

Shireen and her older brother Hasan became activists the day a little boy approached them with a surprising request. Instead of asking for money, he said, "Will you teach me?"

In Shireen and Hasan's part of the world, many parents are too poor to send their children to school. Instead, young children are forced to work or beg for money on the streets.

Hasan and Shireen decided to devote their free time to helping kids who wanted an education. They created The Street School – the first school of its kind in Pakistan. The school day starts at 4pm so that students can spend their days working and still get an education. There are now two locations serving hundreds of students, and the siblings employ six teachers.

ADDISYN GOSS

"IT GIVES ME HOPE THAT I AM HELPING."

Born: 2007 **From:** USA

Known for: creating Snuggle Sacks for homeless people

Addisyn met her grandfather for the first time when she was eight. He was living at a war veterans' home but had been homeless for a number of years. This startling fact gave Addisyn an ambitious idea. What if she made survival kits for homeless people?

She put out a request for donated supplies and assembled fifty Snuggle Sacks. Donations kept pouring in, especially after she went on a local radio station to talk about her mission. Friends and family helped deliver the sacks to nearby shelters and soup kitchens.

Since 2015, Addisyn has given away more than a thousand Snuggle Sacks. Each one is filled with toiletries, snacks, bottled water, a blanket and something warm to wear. There's also a personal note from Addisyn that includes this quote: "Never stop believing in hope, because miracles happen every day."

EIGHT WAYS TO HELP HOMELESS PEOPLE

People tend to think of those in need during the holidays, but homeless people need support and kindness all year round. Here are some ways you can help:

1. **Organize a food or clothing collection.** Gather a group of friends to collect food or clothing in your community and donate items to a homeless shelter.

2. **Have a chat.** If you offer a homeless person food or a hot drink and they don't want it, don't be offended. Try spending a few moments talking to them instead.

3. **Give away small things.** There are other things you can give to a homeless person besides money, like a pair of warm socks, hand warmers, gloves, a hat or a book.

4. **Remember furry friends.** Homeless people often have a four-legged friend as their only companion. Both owner and pet may be grateful for some dog food, cat treats or a soft blanket to curl up on.

KEEP IN MIND

If you see a homeless person and want to help them, remember that the usual rules for staying safe with strangers apply – the best thing to do is to ask a parent or another trusted grown-up to come with you.

It's also worth bearing in mind that you should ask the person if they want (or need) something before you hand it over. If your gift is refused for any reason, just smile and continue on your way!

5. **Smile!** Homeless people often feel ashamed or lonely. A smile and a kind word can make a big difference, especially when someone's having a tough day.

6. **Be ready to help.** Carry cereal bars in your pockets or bag to give to anyone who could use a snack. A bottle of water can be welcome, too. Ditto for those small bottles of hand lotion you get at hotels. Ask your parents or other adults to collect some for you!

7. **Help during any season.** Check your closet for any winter gear you don't use any more, like a pair of gloves, a scarf or a knitted hat. Carry one at a time and give it away to someone in need. In the summer, a clean T-shirt can mean a lot for someone living outside in sweltering heat.

8. **Be compassionate.** Every person living on the streets has his or her own story. No matter what made them homeless, they all deserve love and kindness, even if they don't want your help.

KIDS HAVE RIGHTS, TOO

It's great being a young person, right? That's what grown-ups often say. But it's a mistake to think that children have it easy all the time. In fact, it's so tough for some that they don't really have a childhood. Instead of play dates and parties and pets and other fun stuff, they might…

sleep in temporary accommodation, a car or even on the streets

be abused or forced to work hard all day

have to get married at a young age

regularly go without food

never get a chance to go to school

suffer from disease or poor health

Children deserve better than this. **A LOT** better. That's why world leaders got together and created a special treaty called the United Nations Convention on the Rights of the Child. Almost 200 countries have signed it, making it the most widely accepted human rights treaty in history. It sets out four basic rights that every child should have – this means that they are your rights, too:

Every child has the right to exist. This means food, shelter and healthcare and having their name registered as a citizen of their country.

Every child has the right to reach their full potential. This includes having an education as well as a chance to play and just be a kid.

Every child has the right to be protected. Children should never be used, abused or neglected, forced to work or harmed in any way. Every child deserves a loving and secure family, too.

Every child has the right to freedom of thought. They should be involved in decisions that affect their life and allowed to have their own ideas and opinions.

Children's rights activists work on behalf of young people everywhere. They try to raise awareness about children's living conditions and encourage governments to stick to the terms of the treaty. Some of these activists are still young people themselves!

NOT COOL

Although the **USA** signed the United Nations treaty on **children's rights** it has **never made it official**. It is the **only** UN member state that still hasn't done so.

KIDS TAKING ACTION

"IF PARENTS ARE LOVING AND CARING, CHILDREN WILL HAVE A BETTER ENVIRONMENT TO GROW."

PAYAL JANGID

Born: 2001 **From:** India

Known for: campaigning against child marriage

Payal's parents followed local tradition and arranged for her to get married when she was just eleven years old, but she said NO WAY! When they backed down, she made it her mission to help protect the rights of other kids. She planned rallies and organized protests and became the head of the child parliament in her village, which allows young people to have a voice about matters that affect their own lives.

Payal had a chance to meet US President Barack Obama and his wife, Michelle, when they visited India in 2015. Payal called herself the luckiest girl in India after Michelle gave her a big hug! In September 2019, Payal became the first Indian to be awarded The Changemaker Award from the Bill and Melinda Gates Foundation.

"IT IS UP TO OUR GENERATION TO REMIND OUR LEADERS TO TAKE ACTION FOR CHILD RIGHTS NOW."

SONAM WANGCHEN

Born: 2002 **From:** Bhutan

Known for: making cyberspace safer for kids

Sonam is a popular singer and songwriter who sings mostly in Dzongkha, a language spoken in Bhutan. He is so beloved by the younger generation that he was the perfect choice to become Bhutan's first-ever National Youth Ambassador for UNICEF.

Unlike Sonam's country, the digital world has no borders. More and more young people are spending time online and Sonam wants to make sure it's a safe place for everyone and encourages young people to study and follow their dreams. Sometimes he posts helpful tips on social media, like this one: never share your passwords with anyone!

KIM YOON-SONG

Born: 2003 **From:** South Korea

Known for: fighting for youth voting rights

"HAVING THE RIGHT TO VOTE IS HAVING THE RIGHT TO SPEAK."

Yoon-Song shaved her head in public to protest her country's voting age, which is set at nineteen. Not long ago, she left her home and school after suffering many years of abuse at the hands of her family and teachers. The smallest mistake would be cause for punishment, anything from expressing her opinion to not drinking her milk quickly enough.

Yoon-Song and other activists believe that Korean society ignores, mistreats and discriminates against young people. This may be one reason why South Korea has one of the world's highest rates of youth suicide. Yoon-Song hopes that being allowed to vote at age sixteen will cut down on abuse and give teens more of a say about their own lives.

MILLIE BOBBY BROWN

Born: 2004 **From:** England

Known for: using her fame to help other kids

Millie is an Emmy Award-nominated actor but her life hasn't always been smooth sailing. As a young girl, she was so tormented by 'soul-destroying' bullies that she had to change schools. This triggered huge anxiety but was also her motivation to become an activist.

Millie became the youngest ever UNICEF Goodwill Ambassador aged fourteen and was named one of the world's 100 most influential people by *Time* magazine. At the 2018 Kids' Choice Awards, she made a plea for gun control, then turned to show the names of all seventeen victims of the Marjory Stoneman Douglas School shootings on the back of her shirt.

Millie uses social media to talk about the issues she cares about most, like girls' education, climate change and cyber-bullying. She wants to make sure that kids know their rights!

"THE WORLD CAN BE A SCARY PLACE SOMETIMES, BUT IT'S IMPORTANT TO REMAIN HOPEFUL AND TO BE KIND TO OTHERS."

STORYTELLING TIPS

What do all the young activists in this book have in common? They all know how to talk to people and tell a good story. Whether it's giving a presentation, doing a media interview, rallying a crowd at a protest or strategising with other activists, the same basic principles apply:

Have a point. First things first: what's your message? You should always have at least one main point, such as, 'Avoiding plastic helps save wildlife'. Stay focused – having too many messages is just as bad as having none at all.

Tell a story. A good story needs a beginning, a middle and an end – and so does a good speech or interview. For example, you might start off with what got you interested in a certain cause, then talk about how you learned more about it and conclude with what you're doing to take action.

Connect. Even though you're doing most of the talking, it's best to think of a speech as two-way conversation. Keep your audience in mind and use language they'll understand.

Practise, practise, practise! Public speaking is a lot like riding a bike – it's a skill that takes practice. Test-drive your speech on a few friends or in front of a mirror or on video. The more you do it, the more confident and comfortable you'll feel.

Keep it short. People often have short attention spans so be as concise as possible!

ONE YOUNG WORLD

GENERATION HOPE is the most informed, educated and connected generation in human history. One Young World is a global forum for young world leaders. Every year its annual summit brings together the brightest young minds from more than 190 countries.

MAYBE YOU'LL BE INVITED ALONG NEXT YEAR?

"WE ARE STRIKING TODAY BECAUSE WE HAVE DONE OUR HOMEWORK AND THEY HAVE NOT."

GRETA THUNBERG

RADICAL KINDNESS

Radical kindness is going out of your way to be kind to everyone, even people you don't know. It's also about treating people with kindness even when they aren't nice in return. That's what makes it so 'radical'. Think of it as the opposite of being a bully.

No matter your age or who you are or where you live, you can become a 'compassion activist'. You have the power to transform your school, your neighbourhood and your community to make them better places for everyone. Places that are kinder, friendlier and more accepting.

One good example of radical kindness is 'The Starfish Story'. Ever heard it?

A man was walking along the beach one day when he noticed a boy picking something up and putting it into the ocean. He asked, "What are you doing?" The boy replied, "Helping put starfish back into the water. If I don't, they'll die." The man laughed. "There are hundreds of starfish on this beach! You can't possibly make a difference." The boy listened politely, then picked up another starfish and threw it into the sea. He smiled and said, **"I made a difference for that one."**

It's official. Caring is cool. None of us can ever know exactly what anyone else is going through in life. One small act of kindness might make a huge difference to them. That's why radical kindness can be just as powerful and important as other forms of activism, like **#FridaysForFuture** school strikes.

At a time when news headlines focus more on unkind acts than kind ones, we need people like you to put positive energy into the world. Now more than ever.

IN A WORLD WHERE YOU

CAN BE ANYTHING,

BE KIND

KIDS TAKING ACTION

CODY MCMANUS

Born: 2010 **From:** Scotland

Known for: collecting toys for children living in poverty

"I WANT TO KEEP IT GOING AS LONG AS POSSIBLE AND HELP AS MANY CHILDREN AS I CAN."

Cody loves Christmas and was shocked to learn that some children don't get any presents. He decided to make a difference by organizing Cody's Christmas Box. Cody is a keen boxer and his goal was to fill an entire boxing ring with donated toys and gifts. The campaign was a huge success, with over 2,000 toys donated.

Cody didn't stop there. He still writes hundreds of letters asking people and companies to help him help others and his Christmas Box campaign gets bigger every year. He also raises money by undertaking other challenges, such as cycling 58 km (36 miles) from Edinburgh to Falkirk. In recognition of his selfless efforts, Cody earned the Diana Award in 2018.

CAMPBELL REMESS

Born: 2004 **From:** Australia

Known for: making comfort bears for kids

"BEING KIND AND NOT MEAN WILL CHANGE THE WORLD A LOT."

When Campbell was nine, he wanted to buy some toys for kids spending Christmas in hospital. But he has eight brothers and sisters so his parents couldn't afford to do that. Campbell took matters into his own hands. He found some patterns for teddy bears and taught himself to use his mum's sewing machine.

Campbell has now made and given away too many bears to count. He has started his own charity, Project 365, with a goal to create one bear a day for an entire year. Now his comfort bears go to families with sick children and people who have cancer. Some of his bears have even gone overseas!

FAITH DICKINSON

Born: 2003 **From:** Canada

Known for: creating Cuddles for Cancer

"BEING KIND DOESN'T COST A THING! EVEN A SMALL ACT OF KINDNESS CAN MEAN A LOT TO SOMEONE."

When Faith was nine, her aunt had cancer. Faith made her a soft cosy blanket that her aunt loved – it gave her comfort and kept her warm during her medical treatment. Faith wanted others to feel loved and special too, so she came up with 'Cuddles for Cancer'.

Today, Faith has made over 4,000 cuddle blankets and sent them to more than thirty countries around the world. At first, the blankets were made solely for people with cancer, but now they go to people with other illnesses and diseases as well.

Military veterans and soldiers also receive care packages with blankets, which are a heartfelt reminder that someone cares. And, as Faith likes to say, "Everyone deserves a cuddle." She earned the Diana Award in 2017 and was also invited to the wedding of the Duke and Duchess of Sussex.

LIAM HANNON

Born: 2007 **From:** USA

Known for: founding Liam's Lunches of Love

"START SMALL, START LOCAL. DO SOMETHING THAT YOU LOVE AND THAT YOU THINK WILL MAKE A DIFFERENCE."

It all started when Liam's dad told him that if he didn't want to go to summer camp, he had to find something productive to do instead. So, Liam came up with a way to spread some love to the homeless people he saw in his neighbourhood. He launched Liam's Lunches of Love with twenty brown paper bag lunches. Each bag had a peanut butter and jam sandwich, a snack, a piece of fruit and a bottle of water. Liam also wrote a cheerful message on every bag.

The lunches were such a hit that Liam soon needed a bigger cart and launched a GoFundMe campaign to buy one. His solar-powered cart now serves up to seventy hot meals a day. Liam has given away more than 4,000 free lunches and he has no plans to stop now. His 2019 Million Lunch Challenge inspired many others to join the fight against hunger, too!

50 RANDOM ACTS OF KINDNESS

1. At school, if you see someone sitting on their own, ask them to join you.
2. Say something nice about someone.
3. Write a small note of thanks to someone who helped you.
4. Wash the dishes or take out the rubbish without being asked.
5. Point out when someone's shoelace is undone or their backpack is open.
6. Greet your neighbours if you know them well.
7. Say hi to a classmate you don't normally talk to.
8. Donate books that you no longer read to a charity shop or hospital.
9. Let your brother or sister choose which TV show to watch.
10. Speak up for someone who is being bullied at school.
11. Send a card to a distant relative.
12. Bake cookies for your teacher.
13. Donate toys to a homeless shelter.
14. Smile at someone, just because.
15. Thank the bus driver.
16. Help your neighbour with their wheelie bin.
17. Tell a joke to brighten someone's day.
18. Return the trolley at the supermarket for your parent.
19. Ask your mum or dad how their day went.
20. Pick up litter.
21. Make a bird feeder for your garden.

22. Forgive someone who did something you didn't like.

23. Hold the door open for someone.

24. Carry an extra cereal bar to give to a homeless person.

25. Clean your room without being asked.

26. Write a positive message on the pavement with chalk.

27. Rake leaves or shovel snow for an elderly neighbour.

28. Read a story to your pet.

29. Start a Little Free Library outside your house.

30. Compliment someone on their outfit.

31. Sign up to be a volunteer.

32. Write a poem for a friend.

33. Make a care package for a soldier or a homeless person.

34. Collect money for your favourite charity.

35. If you have a skill, teach a friend how to do it.

36. Draw a flower for a friend.

37. Let your sibling be Player 1.

38. Look after a neighbour's pet when they're away.

39. Collect canned goods for a food bank.

40. Surprise your family by making a dessert.

41. Make a get well card.

42. Get out of bed on a school day without complaining.

43. Give someone you care about a hug.

44. Visit someone in a care home.

45. Leave a friendly note in a library book.

46. Read a story to a younger child.

47. Give someone a high-five.

48. Post a nice comment on social media.

49. Remember to say please and thank you.

50. Tell someone you love them.

WILDLIFE SAVERS

TODAY, WE ARE LOSING ANIMAL SPECIES FASTER THAN WE ARE DISCOVERING NEW ONES. IT'S SAID THAT 60 PER CENT OF THE WORLD'S ANIMALS HAVE VANISHED IN LESS THAN FIFTY YEARS. AT THIS RATE, A MILLION SPECIES WILL BE GONE WITHIN JUST A FEW DECADES. WHO'S TO BLAME? WE ARE. YOUNG PEOPLE ARE THE NEXT GENERATION OF ZOOLOGISTS, ENVIRONMENTALISTS, BIOLOGISTS AND CONSERVATIONISTS, AND MANY ARE TAKING ACTION TO PROTECT WILDLIFE NOW, BEFORE IT'S TOO LATE.

"WHAT YOU DO MAKES A DIFFERENCE, AND YOU HAVE TO DECIDE WHAT KIND OF DIFFERENCE YOU WANT TO MAKE."

Jane Goodall

WILDLIFE ACTIVISTS TEND TO FOCUS THEIR EFFORTS ON FOUR MAIN AREAS:

Selfies with exotic animals. Many animals – even endangered species – are snatched from their natural habitats and sold as props to be held, stroked and cuddled for selfies. Tourists often have no idea that their brief snapshot comes with a high price – a lifetime of cruelty for an innocent creature.

Endangered species. Since 2000, the West African Black Rhino, the Caribbean Monk Seal, the Pyrenean Ibex and other species have gone extinct. It's possible that ALL species of rhino will be gone for good within your lifetime. Countless other animals risk the same fate if we don't act now.

Loss of wildlife habitat and resources. Climate change and cultivating things like palm oil cause massive deforestation, leaving animals no place to go.

Pollution. Millions of animals on land and in the water have died because of our pollution. One study found that 90 per cent of all seabirds eat plastic rubbish, killing a million of them every year.

We can't turn back the clock and bring back what is now lost forever. It's time to take action to make sure that no more damage is done!

KIDS TAKING ACTION

BELLA LACK

Born: 2003 **From:** England

Known for: raising awareness about wildlife

"WE NEED TO VIEW NATURE AS SOMETHING TO PROTECT RATHER THAN EXPLOIT."

Bella has been fascinated with wildlife and animals from a young age. Her life changed aged eleven when she watched a video about palm oil farming and the toll it took on orangutans. So, she overcame her shyness and gave a talk about orangutans at school.

Her activism took off from there, with Bella setting up Twitter and Instagram accounts to empower others with information about wildlife. Her social media platform grew and grew, so much so that she was asked to serve as a youth ambassador for the Jane Goodall Foundation, the RSPCA, the Ivory Alliance and the Born Free Foundation. Now Bella's mission in life is to protect wild animals and promote conservation efforts.

ANYA DE SARAM-LARSSEN

Born: 2005 **From:** Sri Lanka

Known for: starting the elefriendly bus

"I WANT FUTURE GENERATIONS TO BE ABLE TO LIVE TOGETHER WITH THESE GIANT, BEAUTIFUL CREATURES."

In rural Sri Lanka, children sometimes have to miss class because of danger from elephants on their walk to school. The elephants can be aggressive, especially when they are hungry and searching for food. The on-going conflict results in many deaths each year, humans and elephants alike.

Anya came up with a solution – an 'elefriendly' bus! She started a fundraising campaign to buy a vehicle that could safely transport school children through an area known for dangerous encounters with elephants. It worked! Anya was named a 2018 International Young Eco-Hero by Action for Nature.

LIBBY FISHER

"NO VOICE IS TOO LITTLE, NO HANDS ARE TOO SMALL TO HELP SAVE OUR WILDLIFE."

Born: 2005 **From:** Australia

Known for: founding Libby's Koala Crusade

Libby first fell in love with koalas on a visit to a zoo. She learned that these teddy bear-like creatures are an endangered species, mostly due to habitat loss. Heatwaves and drought have taken a toll on koalas' leafy diet, forcing them to come down from the trees and be exposed to predators and road traffic.

Libby was so saddened by this that she started collecting money for charities that are working to save koalas. So far, she has raised more than AU$30,000 (£16,000) for various rescue groups, zoos and wildlife hospitals. Her work is needed more than ever after Australia's devastating bushfires in 2019, which killed hundreds of koalas and burned most of their habitat.

HUNTER MITCHELL

"KEEPING ANIMALS SAFE FROM EXTINCTION IS GOING TO NEED WARRIORS. I WILL ALWAYS BE ONE OF THOSE WARRIORS."

Born: 2008 **From:** South Africa

Known for: founding Save the Baby Rhino

When Hunter heard that a baby rhino had been abandoned by its mother, he took the news hard. Rhinos are endangered and he knew how important it was for the little orphan to survive. To help pay for its care, he donated his own pocket money and asked others to give, too. Hunter's mission quickly expanded to spreading the word about the plight of all rhinos.

So far, he has raised over R250,000 (£13,000) to save the species and train anti-poaching teams. He also makes educational videos, gives speeches and does hands-on work at a wildlife sanctuary. The rhinos now recognize him and come running over to greet him! Hunter was named an International Young Eco-Hero in 2018.

DID YOU KNOW?

10 to 15 per cent of earth's known species on land are found in the Amazon Rainforest, including 40,000 types of plants, 1,300 species of bird and 2.5 million insects. **And no one knows how many more are yet to be discovered!**

GO WILD FOR WILDLIFE!

1. **Be respectful when observing or taking photos of wildlife.** Do not get too close and never chase or tease them.

2. **Feed with caution.** Feeding wildlife can make them less fearful of and dependent on humans, which can put their safety and survival at risk. Bird feeders are okay but don't feed bread to ducks, swans or geese.

3. **Stick up for 'scary' animals.** Some people are so afraid of animals like bats, wolves or sharks that they try to harm them. But these animals all deserve to live and also play an important role in their ecosystem.

4. **Leave no trace whenever you go camping, hiking, boating or on a picnic.** Especially plastic rubbish that can harm wildlife.

5. **Plant wildflowers in your garden to attract bees.**

6. **Ask your parents to avoid toxic pesticides or fertilisers in the garden.** Some grown-ups may need a bit of convincing to change their ways. Be polite and well prepared with facts and it will be much easier to get others on board!

7. **Ban balloons.** Never release them into the air. Balloons are dangerous for seabirds, sea turtles and other wildlife.

8. **Shop compassionately.** Avoid anything made from animal parts, such as scales, feathers, tortoise shells, seashells, coral or ivory.

9. **Skip the selfie.** Riding elephants, holding monkeys and snaps of lion cubs might seem harmless but it often isn't. Even if they are bred for that purpose, they are not being allowed to show their natural behaviours or live in their natural environment.

10. **Lights off!** Bright lights can confuse migrating birds at night, causing them to fly around in circles until they are totally exhausted. Turn off any unnecessary lights at night in the spring or autumn when birds are migrating. It saves electricity too!

11. **Look, don't touch.** When observing wildlife, it's best to do so at a safe distance. You won't disturb animals this way and they have better things to do than hide from you – like looking for food and caring for their young! At the beach, avoid touching or moving plants, coral or other animals, as it could harm them.

12. **Wear reef-safe sunscreen.** An estimated 14,000 tons of the stuff ends up in oceans every year. Use biodegradable products made without ingredients that harm marine life.

13. **Eat only sustainable seafood.** Overfishing means some fish are in danger of extinction. If you eat fish, make sure it's ethically sourced.

14. **Don't pour harmful chemicals like household cleaners down the drain.** Toxins can end up in waterways and oceans and hurt fish, dolphins and other sea creatures.

15. **Avoid places featuring animal 'entertainment', like circuses, elephant rides and dolphin shows.** They all cause animals a lifetime of captivity and suffering.

16. **Sponsor an endangered animal.** The World Wildlife Fund and other charities offer sponsorships as gifts.

17. **Sign petitions** to show decision-makers you support taking action on a wildlife-related issue.

18. Watch *Blue Planet II* and other documentaries on nature and wildlife.

19. **Bid adieu to zoos.** Although some zoos do important conservation work and research, it can still be very disturbing to see animals in captivity. Instead, save your ticket money and donate it to an organization working to safeguard animals in their natural habitats.

20. **Go whale watching from afar.** Whale watching has to be done carefully – it can be distressing for the animals if boatloads of people get too close. Only go with a respectable tour company that cares about the whales' well being.

IT'S NOT EASY BEING GREEN

We only have one planet and we are treating it rather badly. Sometimes, the people sticking up for our world are also treated badly, especially young activists.

The truth is, not everyone wants to face facts because it means they must change their way of life. Ask them to give up burgers or bottled water and they push back. Others refuse to admit that global warming is happening or say it's not that serious or insist that it's all a government plot. **Umm, no!**

If being in denial isn't bad enough, people often say mean things about anyone who dares to speak the truth. Hello?! The people working hard to help the planet aren't the problem – it's the sceptical ones who try to shut them up. Throwing insults around won't stop climate change, nor will staying quiet or quitting.

"WE MUST FOCUS ON WHAT WE CAN DO, NOT WHAT WE CAN'T DO."

Greta Thunberg

Making the world a better place takes more than passion and guts. Here are a few things every young activist should know:

BE INFORMED.

Read up on the issues before diving into the fight. Knowledge is power!

NEVER GIVE UP.

Being an activist isn't easy at any age, but it's important to fight for what you believe in.

THINK FOR YOURSELF.

Don't automatically believe everything you hear or read (especially stuff posted online). Stick to reputable sources (ask your parents and teachers for advice) and always check the facts on more than one website.

LEAD BY EXAMPLE.

Inspire others with your actions and your desire to create a better world.

BE BRAVE.

Prepare for pushback and never let anyone say you can't do something. There may be some resisting but it's worth persisting.

GET ORGANIZED.

Being disorganized wastes time and activists don't have time to waste!

STAY POSITIVE.

Remember, even if it looks impossible, there is always a way.

YOUTH IS A SUPERPOWER.

You have stamina, energy and enthusiasm. All good things!

KNOW WHEN TO LISTEN.

Listening to others is as important as listening to yourself.

DON'T BE AFRAID TO START SMALL.

The important thing is to start!

KIDS TAKING ACTION

"IT IS OUR FUTURE AND OUR ONLY PLANET. I WILL STILL KEEP ON FIGHTING."

HONGYI 'HOWEY' OU

Born: 2003 **From:** China

Known for: being China's first student striker

China has more people and creates more greenhouse gases than anywhere else on earth. Kids there aren't taught much about the environment, but that doesn't mean they aren't concerned. After Howey learned about ocean pollution and the #FridaysForFuture movement, she wanted to go on strike, too. But there was a problem. A BIG problem. Protesting is illegal in China and Howey's friends and parents feared she would be arrested.

She bravely went ahead and announced her strike on Twitter anyway, then sat outside a government building with a hand-lettered sign. Her solo protest ended after a week, when she was hauled off to a police station and questioned for four hours. Released at last, she became more determined than ever to raise awareness and get the Chinese government to do something about the climate crisis.

"I REFUSE TO WAIT UNTIL I'M 'OLD ENOUGH' OR 'SMART ENOUGH.' I WILL START NOW."

MATILDE BONDO DYDENSBORG

Born: 2004 **From:** Denmark

Known for: being a youth congress delegate

After Matilde first heard about climate change, she'd sit in class each day and wonder what she was doing there. The world was in a state of emergency! She did some research and learned about student strikes for the climate. Matilde asked all her friends if they wanted to protest with her. They agreed, but only because they wanted to skip class, not because they wanted to demand government action. So, Matilde founded an eco-council at school to show that an earth-friendly lifestyle isn't difficult. She is now a delegate at the International Congress of Youth Voices.

INGA ZASOWSKA

Born: 2006 **From:** Poland

Known for: solo striking in Poland

"[POLITICIANS] DO NOT THINK ABOUT THE IMPACT OF BURNING FOSSIL FUEL. THEY DO NOT THINK ABOUT THE FUTURE OF YOUNG PEOPLE."

Poland has some of the most polluted air in Europe. It's also experiencing the effects of climate change, with droughts, fires and extreme heat. Inga grew so concerned that she decided to speak out by picketing outside the Polish parliament. Every Friday during her school summer holidays, she sat with her mother, holding a sign that read 'Holiday Climate Strike'. Her bravery attracted a lot of positive media attention, but also stirred up debate and many unkind comments.

Some people said Inga was a pawn in a political game. Others put up fake Twitter and Facebook pages and pretended to be her. Some even claimed that her schoolgirl hairstyle was a trick to win people's sympathy! Inga's comrade-in-plaits Greta Thunberg came to her defence on Facebook, calling her "a true hero, standing up for everyone's future". Very true!

"OPENING THE EYES OF PEOPLE WHO ARE BLIND TO REALITY IS WHAT I NEED TO WORK ON IN JAPAN."

ISAO SAKAI

Born: 2002 **From:** Japan

Known for: speaking out against climate change

Until very recently, Isao didn't know much about climate change and didn't really care about global warming, either. Neither did his friends, or most people in Japan, for that matter. But that all changed when he went to study abroad in the US.

Isao took an environmental science class and it was quite an eye-opener. He finally realized that if no one does anything, global warming would become unstoppable. His next thought was, if somebody has to deal with it, why not him?

That's how Isao became an activist, not an easy thing to be in his culture. Activists must be willing to fight back and speak out, but in Japanese culture standing out from the crowd is frowned upon – there's even a saying: "The nail that sticks out gets hammered down." Isao knows it will be a big challenge to rally Japanese kids and get them to attend #FridaysForFuture protests. Nevertheless, he is determined to succeed.

HOW TO DEAL WITH ECO-MUGGLES

Ever meet someone who doesn't think climate change is real? It's hard to believe but there are still many people who refuse to accept scientific facts and admit what's happening. When it comes to things like global warming, these are the **'eco-muggles'** of the activist world. They are like the muggles of Harry Potter lore, who lack any sort of magical ability.

Young activists are often told they don't know what they are talking about. They hear comments like, "Why should anyone believe you?" or "What you're saying is ridiculous!" The best way to respond is to stay calm and polite (as hard as that might be). **Here are four ways to respond to eco-muggle smack-downs:**

"THIS ISN'T ABOUT ME, IT'S ABOUT_______________ [THE FACTS]."

"MY AGE DOESN'T CHANGE THE FACT THAT SOMETHING IS WRONG AND NEEDS TO BE FIXED."

"I'M WELL AWARE OF THE PROBLEM AND WANT TO TALK ABOUT POSSIBLE SOLUTIONS."

"I MAY BE YOUNG BUT EVERYONE DESERVES TO BE TREATED WITH RESPECT."

DON'T FEED THE TROLLS

Haters and trolls are people who go online – often anonymously – and make mean comments. Unfortunately, a lot of young activists get pushback from people like this. Don't let that stop you from making the world a better place! Here are some tips for how you (and your parents) can deal with them:

Don't respond. Trolls are looking for a reaction – so don't give it to them. Taking their bait only encourages them to be nasty. When they don't get a response, they'll go away.

Ignore them. Focus on the positive comments and move on.

Laugh at them. Having a sense of humour can really help. Greta Thunberg says she laughs at trolls on Twitter: "Some nights I have a competition with myself to find the most absurd."

Delete, block or unfollow. Be sure to delete mean comments as soon as you see them. If it feels necessary, take a screenshot first. Then ban or block users online, which is easy to do.

Report it. Let the website know if someone is trolling you. Facebook and Twitter can suspend the accounts of trolls.

It's not about you. Chances are that the attackers surfed the net looking for someone to pick on. Who better than a kid?! How pathetic. Try not to let a troll's words hurt your feelings.

Tell the police. If the abuse is serious, tell an adult and report it to the police. Online abuse and harassment, known as 'trolling', is against the law.

Be mindful of what you are posting. Try not to post any personal information about yourself, like where you live. If you wouldn't be happy sticking the information up at school for everyone to read, don't put it online!

READY STEADY ACTIVATE!

HERE ARE SOME WEBSITES, BOOKS AND MAGAZINES THAT YOU CAN USE TO FIND OUT MORE.

WHEN USING THE INTERNET, PLEASE REMEMBER THE GOLDEN RULES OF ONLINE LIFE:

- You need to be at least thirteen to use social media.
- Keep your location and personal information private.
- Be smart – don't agree to meet face-to-face with an online friend, or send them photos of yourself, until you've spoken to an adult you trust.
- Report anything abusive or that makes you feel uncomfortable to a trusted adult.
- Remember your digital footprint – everything you post online is permanent.

ACTIVISTS' SITES

Addisyn Goss snugglesacks.org

Alex White appletonwildlifediary.wordpress.com

Amanda Southworth withastra.org

Anuna de Wever twitter.com/anunadewever

Autumn Peltier www.facebook.com/Waterwarrior1

Bana Alabed twitter.com/AlabedBana

Bella Lack callfromthewild.com

Biljana Stojkovic 'I Create My Life' on YouTube

Campbell Remess project365.org

Claire and Sarah Jordan sarahandclairesfooddrive.com

Dara McAnulty youngfermanaghnaturalist.wordpress.com

David Wicker twitter.com/davidwicker_hf

Desmond Napoles desmondisamazing.com

Faith Dickinson facebook.com/CuddlesForCancer/

Fionn Ferreira fionnferreira.com

Genesis Butler genesisforanimals.org

Greta Thunberg facebook.com/gretathunbergsweden and twitter.com/GretaThunberg

Hannah Alper callmehannah.ca

Havana Chapman-Edwards twitter.com/TheTinyDiplomat

Howey Ou facebook.com/OuHowey and twitter.com/howey_ou

Hunter Mitchell facebook.com/Raisethebabyrhino

Inga Zasowska facebook.com/inga.zasowska

Jahkil Jackson officialprojectiam.com

Jamie Margolin thisiszerohour.org

Julie Seven Sage '7 Sage Labs' on YouTube

Kheris Rogers flexininmycomplexion.com and 'Beyond Bullied' on YouTube (SoulPancake)

Liam Hannon liamslove.com

Libby Fisher libbyskoalacrusade.weebly.com

Lilly Platt facebook.com/lillysplasticpickup

Luca Berardi yarhkenya.blogspot.com

Lyla-Rose O'Donovan facebook.com/lylaandlilleysstars

Maanasa Mendu twitter.com/maanasamendu02

Madison Pearl Edwards facebook.com/madibelize

Mari Copeny maricopeny.com

Marley Dias marleydias.com

Max Lawton clearfree.co.nz

Melati and Isabel Wijsen byebyeplasticbags.org

Mikaila Ulmer meandthebees.com

Millie Bobby Brown facebook.com/milliebobbybrown

Nikhiya Shamsher bagsbooksandblessings.com

Nikki Lilly Christou search for 'Nikki Lilly' on YouTube

Oskar Petersen OSKeco.com

Rahul and Rohan Raju facebook.com/rahulrajuandrohanraju

Ryan Hickman ryansrecycling.com

Shireen and Hasan Zafar facebook.com/ThestreetschoolbyHasanShireen

Soffia Correia facebook.com/mcsoffia

Sonam Wangchen facebook.com/sonamwangchenofficial

Thomas Truby facebook.com/therubbishclub

Zoe Rosenberg happyhen.org

BOOKS

A Wild Child's Guide to Endangered Animals Millie Marotta

Charlotte's Web E. B. White

Get Your Boots On Alex White

Glimmer of Hope: How Tragedy Sparked a Movement
March for Our Lives founders

Greta's Story: The Schoolgirl Who Went on Strike to Save the Planet
Valentina Camerini

Kids Fight Plastic Martin Dorey

Marley Dias Gets It Done: and So Can You! Marley Dias

Momentus: Small Acts, Big Change
Hannah Alper

No One is Too Small to Make a Difference Greta Thunberg

The Life-Changing Magic of Tidying Up Marie Kondo

The Survival Guide to Bullying: Written by a Teen Aija Mayrock

MAGAZINES

Eco Kids Planet

National Geographic Kids

Science+Nature

The Week Junior

Aquila

WEBSITES

Action for Nature
actionfornature.org

Born Free
bornfree.org.uk

The Diana Award
diana-award.org.uk

DoSomething.org
dosomething.org

Earthchild Project
earthchildproject.org

Fridaysforfuture
fridaysforfuture.org

GirlGuiding
girlguiding.org.uk

Jane Goodall's Roots & Shoots
rootsnshoots.org.uk

Kids Against Plastic
kidsagainstplastic.co.uk

One Young World
oneyoungworld.com

PETA kids
petakids.com

Royal Society for the Prevention of Cruelty to Animals (RSPCA)
rspca.org.uk

Scouts
scouts.org.uk

UK Student Climate Network
ukscn.org

World Wildlife Fund (WWF)
worldwildlife.org

VIDEO/TV

BBC series
Blue Planet II

BBC documentary
The Blue Planet

Ted Talks
ted.com/talks

You are the future.

GLOSSARY

activist – a person who campaigns for some kind of social or environmental change

advocate – someone who publicly campaigns for a particular cause or group

anxiety – feeling worried, nervous or uneasy about something

app – a software application, which can be downloaded to a computer, phone or tablet

Asperger's syndrome – a type of autism, which is a lifelong condition that affects how people communicate with others and how they experience the world around them

atmosphere – the gases that surround earth

biodiversity – the variety of plants and animals living in a certain area

blog – an online journal in which a writer shares their personal views on various subjects

bullying – to repeatedly harm another person, either physically, mentally or emotionally

CO2 – the chemical symbol for carbon dioxide, a colourless and odourless gas

cat colony – a group of stray cats who live together outdoors

climate – the average weather conditions in a certain location over a length of time

climate change – a change in global or regional climate patterns, mostly caused by human activities that have increased levels of CO2 in the atmosphere

coal – a fossil fuel that can be mined from the earth

coding – to translate instructions into a language that a computer can understand

compassion – showing kindness, caring and a willingness to help others

compost – decaying organic material used as a fertilizer for plants

conservation – protecting natural environments and plant and animal life

coral reef – coral reefs look like weird rocks but are actually made up of thousands of tiny animals called polyps

cyber-bullying – bullying that's done online, such as on social media or via text messages

decompose – when materials, plants or animals rot naturally over time

depression – a mental health condition that can make a person feel very sad and lose interest in things they enjoy for weeks or months

discrimination – to treat people unfairly because of their race, disability, religion, age or sex

diversity – a range of different animals, plants, people or things

drought – a period of low rainfall that can cause water shortages and other problems such as crop failures

eco-friendly – something that doesn't harm the environment

ecosystem – a network of living things such as plants and animals and their environment

energy – power that makes things happen, such as heat, light, sound and electricity

environment – an area in which plants, animals and humans live

emissions – the production and release of gases, such as carbon dioxide, into the atmosphere

endangered species – a species at serious risk of dying out forever, known as extinction

entrepreneur – someone who spots a business opportunity and sets up a new business to cater for that market

equality – treating all people the same and giving them the same opportunities

extinction – when a species dies out completely

Extinction Rebellion (XR) – a worldwide environmental movement that uses peaceful protests to get leaders to take action on climate change

factory farming – a system of raising large numbers of livestock (poultry, pigs, cattle), typically indoors and in overcrowded and inhumane conditions

feral – an animal born in the wild or outdoors with little or no human contact, such as a feral cat

fossil fuel – fuels made from animals or plants that died millions of years ago, such as oil, coal and gas

#FridaysForFuture – an international movement of student strikers who take time off from school to demand government action on climate change

fundraising – encouraging others to donate money to support a certain charity or cause

global warming – a gradual increase in temperature in the earth's atmosphere, due to increased levels of carbon dioxide and other pollutants

green – used to describe things that are earth-friendly and don't harm the environment, also see Eco

greenhouse gas – any gas that traps heat, much like a greenhouse in a garden does

habitat – the natural home of plants and animals

humane – to show kindness and compassion to other people and animals. 'Inhumane' means the exact opposite.

LGBTQ+ – an acronym that stands for Lesbian, Gay, Bisexual, Transgender, Queer/Questioning and + means other sexualities

Maori – the indigenous people of New Zealand

marine – related to the sea

mental health – a person's psychological and emotional well-being

methane – a colourless and odourless gas

microchip – a tiny electronic device, which can be inserted under the skin of a pet such as a cat or dog as a form of identification

mogul – an important or powerful person, especially in business

neuter – surgery that prevents a male dog or cat from becoming a father

non-biodegradable – something that cannot break down naturally over time. For example, a banana peel is biodegradable, but a plastic yoghurt pot is not.

non-recyclable – something that cannot be recycled, like most single-use plastic, such as take-out coffee cups

nutrients – the parts of food and drink that a living thing absorbs to stay alive

organism – any living thing, from plants and trees to animals and humans

panic attack – a sudden and overwhelming feeling of anxiety

pescatarian – a person who eats seafood, including fish, but no meat

petition – a formal written request, usually one signed by many people, appealing to authority about a particular cause

poaching – the illegal hunting and capture of wild animals, for example killing rhinos for their horns or elephants for their tusks

pollution – harmful or poisonous substances (such as plastic rubbish or coal dust) that are introduced into the air, water or soil

population – all the people or animals that live in a certain place

protest – to express one's disapproval about something through words or actions

psychology – the scientific study of the human mind

recycle – to collect and reprocess a material so it can be used again as a new product

refugee – someone who has been forced to leave their home due to war, violence, lack of food, a natural disaster or other reason that makes it unsafe to live there

resource – something that is useful to animals or humans, such as food or wood to build new homes

social media – online communities that share information, ideas, personal messages, videos and photos

solar energy – energy from the sun that can be turned into heat or electricity

spay – surgery that prevents a female cat or dog from becoming a mother

species – a type of plant, animal or other living thing

sustainable – using resources in such a way that they will not run out or become hard to find

sustainably sourced – products that are made in a responsible way, not harming the environment and/or treating workers fairly

TED Talk – a video created from a short speech given at a Technology, Entertainment and Design (TED) event

treaty – a formal agreement between countries

trophy hunting – hunting specific types of animals, such as rhinos, lions and bears, so that part of them can be kept and displayed as a 'trophy'. This very cruel practice is now banned in many countries.

undocumented worker – a person who is working in a country without official paperwork – also known as an 'illegal immigrant'. They are vulnerable to exploitation and often paid low wages and work in dangerous conditions.

UNICEF – short for United Nations Children's Fund

United Nations (UN) – a group of 193 different countries that work together to solve worldwide problems

vegan – a person who doesn't eat, wear or use anything that comes from or harms an animal, including milk and cheese, wool and fur, or products that aren't 'cruelty-free'

vegetarian – a person who doesn't eat animals but eats animal products such as milk, eggs or cheese

vigil – when a person or group stays in a place and quietly waits, prays or keeps watch, especially at night

VIP – short for Very Important Person

vlog – a personal website or social media account where a person posts short videos

YOU CAN MAKE A DIFFERENCE

ABOUT THE AUTHOR

KIMBERLIE HAMILTON

Kimberlie Hamilton has written all sorts of things over the years, but what she loves most of all is writing entertaining non-fiction books for kids with curious minds. She believes that everyone and everything has a fascinating story to tell, if one simply knows how to observe and which questions to ask. Kimberlie has two Master's degrees, one in screenwriting from the University of California and another in Cultural & Creative Communication from the University of Aberdeen. She lives with her partner and four cats in Scotland and hopes to have her own animal sanctuary someday!

OTHER TITLES BY
KIMBERLIE HAMILTON!

REBEL DOGS!
HEROIC Tales of TRUSTY HOUNDS
KIMBERLIE HAMILTON

INDEX

D

E

F

G

H

I

J

K

L

M

N

O

P

R

S

T

U

V

W

Y

Z

[illegible]